Overcoming Internet Addiction

by David N. Greenfield, PhD, MS

Overcoming Internet Addiction For Dummies®

Published by: **John Wiley & Sons, Inc.,** 111 River Street, Hoboken, NJ 07030-5774, www.wiley.com

Copyright © 2021 by John Wiley & Sons, Inc., Hoboken, New Jersey

Published simultaneously in Canada

For general information on our other products and services, please contact our Customer Care Department within the U.S. at 877-762-2974, outside the U.S. at 317-572-3993, or fax 317-572-4002. For technical support, please visit https://hub.wiley.com/community/support/dummies.

Wiley publishes in a variety of print and electronic formats and by print-on-demand. Some material included with standard print versions of this book may not be included in e-books or in print-on-demand. If this book refers to media such as a CD or DVD that is not included in the version you purchased, you may download this material at http://booksupport.wiley.com. For more information about Wiley products, visit www.wiley.com.

Library of Congress Control Number: 2021943694

ISBN 978-1-119-71187-2 (pbk); ISBN 978-1-119-71188-9 (ebk); ISBN 978-1-119-71189-6 (ebk)

Contents at a Glance

Table of Contents

CHAPTER 19: **Ten Tips for Overcoming Internet Addiction and Screen Overuse**

Introduction

You can only experience life in the present moment. Internet overuse and addiction interrupts the process of being present in the moment and can rob you of the ability to fully experience living. When you're online, interaction with the world occurs through reflections found on a web page, app, video game, or the latest social media update. The Internet supports the illusion that you are in two places at once, and it appears to create this dissociation, in part, by the Internet's ability to distort time when you're on a screen.

Internet addiction is not new. Although the Internet may be the latest behavior that we discovered to be addictive, the propensity to become addicted to pleasurable substances and behaviors is not unique to Internet use. The structure and function of the reward center of the brain makes you particularly susceptible to the influence of dopamine and the activities that can elevate it — especially when those activities are provided in a variable and unpredictable format.

The brain loves the experience of *maybe*; addiction is about *maybe* finding the same pleasure again, or *maybe* it being even *more* fun. This is not a conscious process, but rather, it happens automatically, thus becoming a neurobiological *chase* for the previous pleasurable hit. The Internet, in a sense, programs you through the pleasure chemical dopamine, and you don't realize you're being conditioned by your screens. Content developers and Internet service providers know these basic behavioral and neuroscience principles, and they use and manipulate them to sell you products and influence what you consume online — including how much time and attention you devote to your screens.

Many people overuse their Internet screens and devices, and end up spending far too much time on them. For some, this use rises to the level of an addiction, where balanced living becomes affected and other psychiatric issues may develop or become pronounced. The Internet is a powerful digital drug, and as such, it must be moderated like any other addictive substance or behavior. This book is intended to help people who may lose their way online and especially for those who become addicted.

About This Book

This book is written in discrete chapters and is set up with multiple independent, but related, topics. Feel free to start wherever you like; some information and key concepts are repeated for emphasis, and refer you back and forth to different chapters. The idea is to immerse yourself and start wherever you feel comfortable. I trust that you'll find the information you need to discover, and because this isn't a novel, each chapter stands alone. If you're unfamiliar with this subject, then jump in and start at Chapter 1; the book is written to build on basic information and concepts, then onto application, and lastly, implementation.

The order of the book presents progressively more interrelated information, such as Chapter 2 on the biology of addiction, Chapter 5 on social media, and Chapter 7 on video games and video game addiction. Part 3 of the book offers information on diagnosis, followed by Part 4, which focuses on living a balanced life with proper screen use.

I want to clarify the perspective from which I write this book. I've conducted research and published in the field of Internet, video game, and technology addiction and its treatment, as well as workplace Internet issues and smartphones and driving, but I still consider myself primarily a clinician. This book is written from a scientifically informed clinical perspective; I've treated or consulted with hundreds of individuals over the last 25 years, many with significant life impacts related to their Internet, video game, and screen use. Numerous references throughout the book will represent my opinions based on these clinical experiences, some of which should be considered anecdotal and case based. I don't represent that everything stated in the book has been scientifically validated, as frankly, we're not quite at that point in the science of the diagnosis and treatment of Internet-related disorders.

A word about addiction: Please note that this is discussed at numerous points throughout the book, but I want to start by clarifying that the word *addiction* is perhaps not the best term *per se.* My approach tends to follow the definition of addiction offered by the American Society of Addiction Medicine, which is deeply clinical in nature. This definitional distinction may seem odd because the word *addiction* is used throughout this text, as well as in the fields of clinical psychology, psychiatry, and addiction medicine; however, I'm using this label throughout the book because it's the term that most of us understand and relate to when describing the complex biopsychosocial phenomenon that produces something we define as *addiction* (see Chapters 1 and 2).

With specific regard to Internet and technology addiction, no scientific agreement currently exists on what the final diagnostic labels will be. It is my best guess, based on the state of the research and clinical trends (including those established by the World Health Organization, the American Psychiatric Association, and the American Psychological Association), that the final diagnoses beyond *Internet gaming disorder* may include a variety of Internet-mediated use disorders that would involve symptoms of abuse and overuse of Internet-based screen technologies. This would be similar to the current diagnostic labeling that we see with alcohol and substance use disorders. Internet use disorders will probably follow suit and will most likely include several diagnoses that overlap or may be related to a specific type of content or Internet portal.

A quick note: Sidebars (shaded boxes of text) dig into the details of a given topic, but they aren't crucial to understanding it. Feel free to read them or skip them. You can also pass over the text accompanied by the Technical Stuff icon, as it offers some interesting but nonessential information about Internet addiction.

Finally, within this book, you may note that some web addresses break across two lines of text. If you're reading this book in print and want to visit one of these web pages, simply key in the web address exactly as it's noted in the text, pretending the line break doesn't exist. If you're reading this as an ebook, you've got it easy — just click the web address to be taken directly to the web page.

Foolish Assumptions

Here are some assumptions about you, the reader, and why you are picking up this book now:

>> You or someone you love may be spending far too much time on one or more screens, and you've noticed the impact that this behavior has on you, on them, or on both of you.

>> You or your loved one are finding the Internet almost too enjoyable (or certainly captivating), and at times you find yourself lost in the bits and bytes of cyberspace — so much so that at times, you lose track of time and also of your larger goals, values, and desires.

>> You want to change your relationship to this useful, but addictive, technology, or you would like to see your children or loved one regain some balance and perspective about their screens. Perhaps surfing, social media, YouTube, pornography, video games, or never putting down their (or your) smartphone has become less tolerable and may even seem harmful at this point.

>> You, just like myself, see the unavoidable pull of the Internet and screens. You recognize that you need to use them, but that perhaps you recognize that you do not need to use them in the manner you currently do, and you'd like to

achieve a more mindful, moderated, and sustainable way to live *with* your Internet screen technologies, as opposed to living *for* them.

>> You know that the Internet and accessible screens are not going anywhere, and that if anything, they will become even more prevalent and intrusive. You also see that this is not all positive and that *too much of a good thing is not a good thing,* and you must exercise your choice and ability to limit your screen use.

Icons Used in This Book

Like all *For Dummies* books, this book features icons to help you navigate the information. Here is what they mean.

REMEMBER

If you take away anything about Internet addiction from this book, it should be the information marked with this icon.

TECHNICAL STUFF

This icon flags information that delves a little broader or deeper than usual into Internet addiction.

TIP

This icon highlights especially helpful advice about understanding, diagnosing, or treating Internet addiction.

WARNING

This icon points out situations and actions to avoid as you work to overcome Internet addiction.

Beyond the Book

In addition to the material in the print or ebook you're reading right now, this product comes with some access-anywhere goodies on the web. Check out the free *Cheat Sheet* for information on Internet addiction basics, reasons why the Internet is addictive, diagnosing and treating an Internet addiction, and living a balanced life that includes the proper amount of screen use.

To get this Cheat Sheet, simply go to www.dummies.com and search for "*Overcoming Internet Addiction For Dummies* Cheat Sheet" in the Search box.

Where to Go from Here

As I mention earlier, you don't have to read this book from cover to cover, but if you're motivated to do so, and you want to have a thorough immersion in the subject, starting from the beginning is a good idea. If you just want to find specific information and then get back to work, look at the table of contents or the index, and then dive into the chapter or section that interests you. Sampling a topic can help you see where you might want to delve in further.

Please remember that there is hope. There is hope for people to change and for addictions to improve and be managed. Also remember that humans are unique in their ability to learn and adapt, and that the brain is neuroplastic and can change, retool, and relearn throughout your lifetime. People have an inherent desire to grow, heal, and improve, and you or your loved one are no exception. That, of course, in no way minimizes the stress and disappointment you experience when your life or your child's life is not where you want it to be. People are capable of great change, and there are many paths to positive improvement, especially with the resources and professional help that are available. We know much more about Internet and technology addiction than we once did, and there are now many more mental health and addictions professionals today, who are trained and experienced in consulting on or treating this problem.

1

Getting Up to Speed on Addiction Basics

Look at the basic factors involved in Internet and technology addiction, and examine ways to overcome overuse and addictive use of your screens. You also find the definition of addiction, and you can check out the similarities and differences between Internet addiction and substance-based and other addictions. Lastly, you discover some of the most overused and abused forms of content on the Internet.

Examine the biological and neurobiological basis of addiction and how the reward centers in the brain are involved in the development and maintenance of an addiction. Addiction is a complex biopsychosocial problem that encompasses numerous aspects of your life.

Recognize that children and teens are perhaps more susceptible to addiction than adults. Numerous hormonal and psychological factors in the developing brain can leave children and adolescents susceptible to Internet and screen-based addictions, as well as addiction in general. Adolescents' unique biological and psychological development provides fertile ground for an addiction.

See why smartphones and the Internet are so addictive. Here you examine the unique characteristics and factors that contribute to the addictive nature of the Internet and the devices that you use to access it. See how smartphones are the *world's smallest slot machines* and how carrying these devices serves as a *portable dopamine pump,* providing intermittent, but unpredictable, pleasurable content.

Chapter **1**

Defining and Overcoming Internet Addiction in a Nutshell

The interesting thing about the word *addiction* is that technically it isn't really a medical term or diagnosis. Although used by nearly everyone, both clinicians and the public, it's more of a popularized term used to describe a set of behaviors or a syndrome. Official diagnostic terms for substance and behavioral addictions include *substance use disorder, alcohol use disorder, pathological gambling,* and *Internet gaming disorder.* For the purposes of this book, I use the term *addiction* for ease and simplicity.

Most people confuse an addiction with physical dependence. Physical dependence occurs when the body gets used to a substance, be it alcohol or drugs. It is characterized by a tolerance to that substance and then withdrawal when the substance is discontinued. Essentially, the body's receptors for that drug become accustomed to having it in the system. When it's no longer available, there are physical and psychological symptoms that we call withdrawal.

REMEMBER

Addiction is more typically defined as a pathological or compulsive use disorder. This means that when you use a substance or engage in a repetitive behavior (such as gambling, Internet use, or video gaming), significant negative effects are created in your life. Despite these negative effects, the user cannot easily stop or may not think they need to stop. This distortion of reality is often inherent to addiction and is also known as denial.

We all engage in pleasurable behaviors and at times take substances that are pleasure inducing. Take alcohol, for instance. Alcohol is a legal psychoactive substance that has long been associated with pleasurable sensations, but unfortunately, it is also known for its addictive potential. Many pleasurable substances and behaviors can produce an addictive response due to their activation of the reward circuitry in the brain.

REMEMBER

There is some confusion over whether *intoxication* and/or *withdrawal* described in alcohol or substance use is also experienced in behavioral addictions such as gambling, food, sex, or the Internet. Clarifying this issue isn't necessary to recognize behavioral addictions, however. Addiction is not simply the intoxication or withdrawal we get from a substance or behavior. It is the creation of a potential set of behaviors and life-impacting consequences reflecting a complex biopsychosocial process. We call it biopsychosocial because it affects our physical health as well as our social and emotional life.

This chapter introduces you to Internet and screen addiction, how to recognize it, and how to get help.

Defining Behavioral Addiction

There is some confusion about what causes an addiction; this confusion often occurs because of the physiological response of tolerance and withdrawal from drugs or alcohol. But what about gambling? With gambling, you aren't ingesting anything, yet you see all the same markers and consequences of addiction, including an impact on social relationships and psychological functioning, as well as on work, legal issues, finances, health, or academic performance.

Gambling addiction (the official medical diagnosis is *pathological gambling*) is part of a group of addictions called *process* or *behavioral addictions;* the American Society of Addiction Medicine, in part, defines addiction as the use of a substance or behavior that causes *negative* and *deleterious* life consequences:

Addiction is characterized by inability to consistently abstain, impairment in behavioral control, craving, diminished recognition of significant problems with one's behaviors and interpersonal relationships, and a dysfunctional emotional response. Like other chronic diseases, addiction often involves cycles of relapse and remission.

REMEMBER

The complex process of addiction almost always involves disruption of reward patterns, motivation, compulsion, executive function, and judgment, but it may not include the physical withdrawal symptoms seen in drug or alcohol dependence. Flip to Chapter 2 for an introduction to the biology of addiction.

Understanding How and Why People Get Addicted to Screens and the Internet

So, what about the Internet? And how exactly do you become addicted to a digital screen connected to the Internet? Well, the answer is not all that different from how you become addicted to other behaviors and substances. Part of what happens with screen use (when linked to the Internet) is that you're accessing content that is stimulating and rewarding to you, but because the delivery mechanism of this digital drug is variable (meaning you don't know *what you will get, when you will get it,* and *how desirable it will be* for you), your brain receives a variably rewarding experience. Each time you get a reward, you receive a small hit of dopamine in your mid-brain (also called the limbic system). This unpredictability, or *maybe* factor, is very resistant to extinction or, putting it another way, addictive.

This is the same way that addiction to gambling works, and that is why I call the Internet the "world's largest slot machine" (see Chapter 3 for more information). Each time you pull the handle on a slot machine or click your mouse, you might win; with gambling, you might win some money — on the Internet, you might win some form of desirable content.

Screen technology is essentially doing the same thing that drugs, alcohol, and gambling do. Obviously, there are differences between the various types of addictions, but the underlying neurobiology is essentially the same. You tend to repeat behaviors that are pleasurable, and the perception of that pleasure is often unconscious and largely biological; the impacts these addictive patterns have are more often behavioral and psychological, but the underpinnings of all addictions are neurobiological.

Many people can experience the excitement and rewards of the Internet and most forms of digital content with few problems, although you will frequently hear people complain about of how *addicted* they feel to their smartphones, Facebook, TikTok, Instagram, Snapchat, Reddit, Twitter, pornography, video games, YouTube, and even streaming sites like Amazon Prime, Hulu, and Netflix. The reality is that some people are unable to limit their use of these forms of content online, and because the Internet delivery system operates on a variable reinforcement schedule, the brain gets used to the *maybe* factor that helps produce the addictive response. The pleasure part of the brain (see Chapter 2) responds with increases in dopamine, and the unpredictability of *what, how much,* and *when* helps create a digital addictive experience. Obviously, the content you might *like* and respond *favorably to* will vary, but the Internet has *so much* available content that it is easy to understand why many people feel addicted to their screens (the reality is that many people are overusing their screens, but may not be addicted).

REMEMBER

It's perhaps important to note here that feeling addicted to your screen may not mean that you are. Many of us are aware of our overuse of our screens, but it may not have risen to a point that meets addiction criteria. Nevertheless, it can be a lifestyle problem that you should address by making some changes in *how* and *when* you use your Internet technology. I talk about signs and symptoms of Internet and screen addiction later in this chapter.

REMEMBER

Addiction represents an extreme problem with a drug or behavior, but it has different levels of severity just like any illness or behavioral health problem. The levels of addiction can range from mild to severe, with each level representing more significant negative impacts on your behavior and functioning. At the lowest level of impact, you might be overusing your technology and screens to a point where you're eating up too much time and energy that might be better spent on other tasks and activities. At the most extreme level, I have seen people whose lives have been severely impacted and limited by their screen use. It is perhaps fair to say that most of us (at times) fall into some level of overuse, abuse, or addiction to Internet technology.

TECHNICAL STUFF

The science and practice of addictionology and addiction medicine define addiction in the same way, whether it involves using substances or engaging in compulsive behaviors. The American Society of Addiction Medicine gives the following more complete definition of addiction:

Addiction is a primary, chronic disease of brain reward, motivation, memory, and related circuitry. Dysfunction in these circuits leads to characteristic biological, psychological, social, and spiritual manifestations. This is reflected in an individual pathologically pursuing reward and/or relief by substance use and other behaviors. Addiction is characterized by inability to consistently abstain, impairment in behavioral control, craving, diminished recognition of significant problems with one's behaviors and interpersonal relationships, and a dysfunctional emotional response. Like other chronic diseases, addiction often involves cycles of relapse and remission. Without treatment or engagement in recovery activities, addiction is progressive and can result in disability or premature death.

Digging into Digital Devices and the Internet

Today, digital screen devices reflect a wide range of technologies and include iPhones and Android smartphones; iPads, Kindles, and other tablets; laptop and desktop computers; and streaming devices and smart TVs. These days, it's very difficult to find something that isn't directly or wirelessly linked to the Internet. You will likely almost always have easy access to an Internet connection portal, which increases your overall risk for developing an addiction or unhealthy Internet use habits.

WARNING

All digital devices have the power to rob you of your time and attention, and this imbalance creates problems.

The power of, and attraction to, the Internet comes from its ability to connect people with other people, information, and services. This has essentially changed the way we live our lives; however, the most powerful aspect of the Internet is its addictive potential. The way the Internet works can lead to increases in the amount of time you spend online, irrespective of the specific content you consume.

REMEMBER

The Internet is neither good nor bad — it's amoral. It has no feelings in its power to captivate you. However, you should always keep in mind that the only goal of everything online is to *capture* and *hold* your *attention*. It's not necessarily a nefarious intention, but nevertheless, it's a potent force whose purpose is to keep you screen-bound (for largely economic reasons). The only way out of the black hole of the Internet is to take back control of your time and attention.

REMEMBER

Smartphones are the world's smallest slot machines. Their ease of access and availability make them highly addictive Internet access portals. And making smartphones even more addictive are the *notifications* they provide you. Each time you receive a notification of some type, your brain registers a triggering signal that a desirable message, information, or content is waiting for you. This facilitates your continually picking up and checking your phone all day long. There are estimates that many of you pick up your phone a hundred times or more a day! The activation of possibly finding something pleasurable when you're checking is *even more* rewarding than the content itself. Nothing is more intoxicating than *maybe*.

See Chapter 4 for more information on the addictiveness of smartphones and the Internet.

Recognizing the Threats

How do you know whether your *digital device* has become your *digital vice?* This isn't an easy question to answer, as everyone has their own personal values about time and everyone is different in terms of how much disruption is tolerable in their lives. Your technology use is often in part determined by your values around how you use your time and the consciousness you bring to your Internet screen use.

The idea that your screen use typically occurs below your conscious radar is well established. *Time distortion* and *dissociation* are common when you are on your screen, so it's very likely that you don't actually know how much time passes when you're staring at your Instagram or Facebook account, or circling down the rabbit hole on Reddit, or caught in an endless YouTube playlist or Netflix binge.

The first step in recognizing your Internet use is to become conscious of how much *content* you're consuming, and to become aware of exactly how much *time* you have unknowingly surrendered to your device. This is perhaps easier said than done, in that most of the time on your device may be spent without a thought, with little awareness that you have begun a slow descent into the electronic sinkhole of the Internet.

The following sections briefly cover addictive platforms and technologies. Part 2 is devoted to these topics.

Social media

What exactly is social media? I explain in Chapter 5 that *social media* is a broad category of applications and websites that are structured around the idea of connecting people, organizations, or businesses around themes, content, or interest areas. Although there are several well-known social media sites, the definition of social media has also expanded to include newer apps and websites that attract users' time and attention. Some examples of social media sites are Facebook, Instagram, LinkedIn, TikTok, Reddit, Snapchat, Vine, Twitter, and YouTube, and even travel sites like Waze and Yelp have a social media component; undoubtedly, countless social applications are integrated into other apps and websites, as businesses have found that *social sells.*

Some social sites are integrated around news and communication, others focus on video and text, and still others are simply about photos and user life updates, posts, and sharing. Some integrate all of these features. Most are without fees to the user; however, none are free. Most accept advertising, and many sell your user data to others for a variety of purposes.

WARNING

Make no mistake: If you think there are no obvious payment of costs connected to social media, then *you* are the payment! Your eyes and attention form the economic engine that drives social media, and many of these platforms have proven to be both addictive and negatively impactful on a variety of psychological levels.

Streaming audio and video

Consuming audio and video content (including music, podcasts, and audiobooks) has become commonplace; in fact, most of what you consume in terms of music, TV, movies, Netflix, Amazon Prime, Hulu, podcasts, YouTube, Kindle, and so on is essentially in streaming format. Even software is streamed or downloaded, and your data is increasingly held in the cloud (on a company's server) instead of on your devices. Streaming means that information and content is pushed to your device in real time, and you watch, read, or listen to it as it is streamed. Sometimes you can download it onto your device and store it for later use.

WARNING

Although no one would argue against the convenience of consuming your digital entertainment in this manner, there are some inherent problems. One major problem is that many of these streaming sites have default settings, called *auto-play,* for audio or video content to keep going to the next movie, TV show, podcast, or YouTube clip — unless you deliberately turn off that feature. The net effect (no pun intended) is that you can end up watching or listening to a lot more content than you intended to or have time for. The automatic "pushed" nature of the content is equivalent to eating out of a large dish with no ability to measure the portion of food (or digital *content*) you are consuming. See Chapter 6 for more about streaming content.

Video games

Video gaming is perhaps the most common reason why people seek treatment in our clinic and residential treatment center. Of all the content areas that are consumed on the Internet, video gaming is perhaps the most problematic that we see in terms of negative life consequences. That doesn't mean that video gaming is bad or inherently dangerous; in fact, most people who use video games have no significant problem with them and are able to use games with little or no negative life impact. In other words, they can use them in a moderated manner. However, there is a small percentage of users (studies suggest between 1 and 10 percent) who cannot self-regulate their use and who spend inordinate amounts of time on gaming platforms, including handheld devices, consoles, and PC-based systems. Most of the patients we treat seem to be having problems with PC-based or console games.

Video games incorporate some very attractive factors that contribute to their addictive potential:

>> They provide stimulating content that is novel, interactive, and dynamic. Games are always evolving, through updates and modifications, to keep the novelty and challenge factor high.

>> When playing a video game, you can experience a level of growing mastery that often creates a sense of accomplishment; you might develop greater efficacy in your skills and a higher ranking in comparison to other users.

>> The game provides a sophisticated variable ratio reinforcement and reward structure (the *maybe* factor), and this structure is modified and changed to maintain user interest and to maximize the dopamine/pleasure response.

All forms of Internet communication facilitate some degree of social connection and group interaction, albeit in a two-dimensional online format, but many video game users find the social component of video gaming to be quite compelling. Many times, gamers are communicating verbally (on a headset) via apps such as Discord or others, and the conversations may not only be about the game being played. Flip to Chapter 7 for more about video games.

Online gambling

Although online gambling is technically illegal in the United States, many sites can be accessed offshore and also found on the dark web (covered later in this chapter). The problem with gambling is that it is potentially quite addictive to begin with, but when you combine this stimulating form of content with an online interface, it becomes even more appealing. Online, there are no thresholds to cross and no one to look in the eye; all you have is unfettered, easily accessible content that is both fun and highly addictive. Online gambling removes the last vestiges of human interaction from the equation in that it's an easy pick-and-click activity with no personal interaction. See Chapter 8 for more information.

Online shopping

Shopping can certainly be addictive, and it can elevate your dopamine levels in the same way that many substances and behaviors do. However, what could possibly be bad about the amazing convenience of shopping online? I do it several times a week and it certainly makes my life much easier.

But the dark side to this convenience is that you're divorced from the impact of your shopping by the ease of being able to search and click and then receive the item within a day or two. Amazon has this down to a science, and they know that the average consumer will *almost always* purchase more than they ordinarily would because of that ease of access and convenience — and by the fact that they bombard you with reminders and images of what you looked at and what you might need or should buy, you're pushed to click some more. This consumer science has the effect of making you less conscious of your purchases, at least until the bill comes. Rarely a month goes by that I am not surprised by how much I have spent at Amazon. Sometimes online shopping can just be a little too easy. Chapter 8 has more details about online shopping.

Online investing

Stock trading and online investing are a perfect match for the Internet. Just like shopping (see the previous section), it allows you ease, convenience, and relative privacy, and it makes trading stocks and other investments simple and near instantaneous.

The problem is that sometimes you need time to think about your investments, and you can easily make an impulsive move online. There is a tendency with all online transactions to end up in a pick-and-click pattern that feels good in the moment but may not always be the best choice. Granted, you can make bad investments offline as well — but again, when you are online, there is no threshold to cross, no one to easily call and speak with on the phone, and no one to run things by. Sometimes all that is fine, and if online stock trading is the end point of a well-researched and thought-out choice, then it can work. But as with online shopping, it is just you and your screen, so there is some potential for higher risk in an online transaction.

Our research shows that users tend to be more uninhibited and impulsive online than with other modalities. When you are online, you are in essence isolated and more apt to make riskier decisions. The addictive component is also important to note here in that each time you buy a stock or make an online trade, you're getting a small hit of dopamine. This occurs irrespective of whether that investment is a good one or not and therefore may be misleading in terms of your judgment. See Chapter 8 for more about online investing.

Online sex and pornography

No discussion of online addiction would be complete without a discussion of pornography. In terms of reasons for people seeking treatment and general complaints about Internet addiction, it is on the top of the list along with video gaming.

In some sense, the pornography and adult entertainment industry contributed to the early development and adoption of the Internet. The porn industry spearheaded some of the first examples of using the Internet to provide online adult entertainment to users and was an early example of effective e-commerce. The ease of access, disinhibition, dissociation, privacy, and perceived anonymity of the Internet makes it a near-perfect medium for pornography, and the fact that there are no boundaries and endless choices further facilitates the potential for addictive use. Of those who use online porn, most report overusing it at times, and a sizable percentage feel addicted to it at least part of the time.

Many other sexual behaviors are also available online that can become addictive, including visiting webcam sites, watching private video shows, engaging in video/phone sex, and websites featuring various form of real-time sex work. In addition, there are numerous hook-up and sex sites that use an online platform to connect buyers and sellers of these services, as well as prostitution. Again, many of these activities existed prior to the Internet, but the Internet is a clear amplifier and facilitator, and makes the use and abuse of these behaviors all too easy. When it comes to sex, you have a double hit in that sex is one of the primary reward drives in the brain, and along with the addictive medium of the Internet, there can be an *amplification* of the addictive risk potential.

The dark web

WARNING

The dark web is an offshoot of the Internet and can offer many things (legal and illegal) that cannot be found on the normal Internet browsers such as Google Chrome, Safari, Firefox, and Microsoft Edge. A plethora of illegal pornography, drugs, and other illicit merchandise and services are available on the dark web, but there are also monitors in place where you can be caught accessing this illegal information or merchandise. Anything and everything is available on the dark web, and it remains the true Wild West of the Internet (and you do not know whom you are dealing with). My general advice is to avoid its use, and if you have any issue with addiction to substances or online pornography, this can be a dangerous place. The dark web *does not* have some of the online checks and balances (as limited as they are) that normal browsers afford.

Identifying the Signs and Symptoms of Internet and Screen Addiction

Following is a general list of things to look out for to determine whether you may be suffering from an addiction to the Internet. Sometimes just developing an awareness of what you're doing can increase your self-consciousness enough to cause you to change your habits and patterns. This is a good place to start. Generally, small changes can be valuable, but you can make those changes only if you are really aware of what you're doing. Chapter 10 has more information on identifying signs and symptoms, and Chapter 11 provides a number of self-assessments.

REMEMBER

Every accomplishment starts with a *goal*, followed by an *assessment* of where you are, and a *plan* for where you want to be:

1. Do you spend more time online on your screen devices (computer, laptop, tablet, smartphone, or smart TV) than you realize?

2. Do you mindlessly pass time on a regular basis by staring at your smartphone, tablet, computer, or smart TV, even when you know there might be better or more productive things to do?

3. Do you seem to lose track of time when on any of your screen devices?

4. Are you spending more time with "virtual friends" as opposed to real people nearby? (Obviously, during the COVID pandemic this is a difficult question.)

5. Has the amount of time you spend on your smartphone or the Internet been increasing?

6. Do you secretly wish you could be a little less wired or connected to your screen devices?

7. Do you regularly sleep with your smartphone under your pillow or next to your bed?

8. Do you find yourself viewing and answering texts, tweets, snaps, posts, comments, likes, IMs, DMs, and emails at all hours of the day and night — even when it means interrupting other things you are doing?

9. Do you text, email, tweet, snap, IM, DM, post, comment, or surf while driving or doing other similar activities that require your focused attention and concentration?

10. Do you at times feel your use of technology decreases your productivity?

11. Do you feel uncomfortable when you accidentally leave your phone or other Internet screen device in your car or at home, if you have no service, or if it is broken?

12. Do you feel reluctant to be without your smartphone or other screen device, even for a short time?

13. When you leave the house, do you typically have your smartphone or other screen device with you?

14. When you eat meals, is your smartphone always part of the table setting?

15. Do you find yourself distracted by your smartphone or other screen devices?

If you answer yes to 50 percent (7 or 8) or more of these questions, then you may want to examine your Internet and screen use.

REMEMBER

Here's an important disclaimer: It should be noted that no medical or psychiatric diagnosis can be made solely from a written test or screening tool. These Internet and screen addiction diagnostic criteria are intended for educational and informational purposes only and are not a substitute for professional medical advice. If you are concerned about your smartphone, Internet, or screen use, you may want to consult with a licensed mental health or addiction professional with expertise in Internet and technology addiction.

REMEMBER

The main thing to look out for is an *overall lack of awareness* of how much time you are spending on your screens. *The more time, the more likely your life will be out of balance.* The content or app is not the most important thing here; rather, it is the amount of *time* you are diverting from balanced real-time living. The power of the Internet, in part, comes from its ability to dissociate you from real life and to become a digital drug by impacting dopamine levels in your brain.

Recovering from Internet and Screen Addiction

You cannot change anything in your life unless you have honest self-appraisal and feedback. The problem with addiction is that often, *self cannot accurately see self*, and people have a great capacity for denial and self-deception when engaging in addictive behaviors that impact their brain reward centers.

Recovery always begins with honest self-evaluation, often with some objective data to help you accurately see what you're doing online. With substance as well as behavioral addiction, there are well-established methods for assessing overuse and life impacts from an addiction. However, there is no simple answer for how much is too much, nor is there an easy fix. As I note throughout this book, addiction is complex and involves mind, body, and spirit — impacting many aspects of your functioning; that said, everyone has a different *bottom line* where they can no longer ignore the fact that their screen addiction is hurting their life in some way. The following sections introduce two options for recovery: self-help and professional help.

TIP

One thing you do have with Internet addiction is a *digital footprint;* that is, everything you do online and on your smartphone can be tracked, and you can see how much time you spend, what websites and apps you use, and what content areas you seem to have a problem with. This feedback can be critical in helping you start the process of recovery by seeing what you're doing, much like keeping a record of the foods you eat when attempting to eat better or lose weight. There are many aftermarket apps and programs that can record, track, block, and monitor your Internet and screen use. Most cellphone manufacturers and service providers have apps that offer a great deal of detailed information on your use. Several companies also produce software that you or an IT professional can install on all your screen devices that can give you accurate and detailed data, which provides you with total usage information and any problem content areas (see Chapter 11 for more options regarding self-help resources).

REMEMBER

Don't be surprised when you look at your usage information and find that it is much greater than you recall it being or that you were aware of. This is normal and is part of that dissociation and time distortion that I talk about earlier in this chapter. It's essential that you get accurate feedback about your use; otherwise, you'll be unable to take control of your screen time.

Exploring self-help options

Self-help options (covered in Chapter 12) have always been a substantive part of any addiction recovery and treatment plan. The most well-known is Alcoholics Anonymous, but there are 12-step and recovery/support programs for nearly every addiction; there are even specific support groups for pornography and sex addiction, including Sex and Love Addicts Anonymous, Sex Addicts Anonymous, and Sexaholics Anonymous. There are also many support and self-help groups for Internet and technology addiction. Groups like Game Quitters, OLG-Anon (On-Line Gamers Anonymous), and others that focus on video gaming and other forms of screen use can be useful, but beware that many of these groups *are themselves* online. Some might argue that this defeats the purpose, but my experience suggests that some help is always better than no help, even if it's online. (COVID also gave us new reliance on the utility of telemedicine mental health and addiction treatment.)

Self-help books and resources can be invaluable in making desired changes in any behavior or addiction. When I first started my work in Internet addiction in the late 1990s, only one self-help book (*Caught in the Net,* published in 1998 by

Dr. Kimberly Young) was available, and my book *Virtual Addiction* was the second, published in 1999. Now, literally dozens of books and resources have been written, and a great deal of medical and scientific research has been conducted on the subject. We know a lot more about this new addiction than we did 25 years ago (see Chapter 12 for more on self-help strategies).

Getting professional help

Sometimes self-help and support groups aren't enough to make the changes you need in your life. Breaking an addiction can be hard, and any behavior change can sometimes be made easier with professional help. My recommendation would be to find a psychologist, psychiatrist, or therapist who has experience in addiction and addiction medicine, preferably with a background in Internet, video game, or screen addiction. Don't be afraid to ask questions regarding their experience (and expertise) in this area, and be wary of any doctor or therapist who underplays the issue and tells you it isn't a real problem.

REMEMBER

For the last 20 years, I supervised psychiatry residents and taught courses on Internet addiction and sexual medicine at the University of Connecticut School of Medicine in the Department of Psychiatry. I have trained many doctors and therapists on how to treat this issue, so please make sure you find someone who has had some training and/or experience in this specialty area. See Chapter 13 for more about finding professional help.

Raising Tech-Healthy Children

At least two generations have now been raised on Internet and screen technologies. I wish I could unequivocally say that this been a good thing. The Internet and digital screens are unquestionably a great innovation and have changed the world in *many* positive ways, but there have been some negative changes as well. There are clear brain and behavioral changes from excessive screen use. It seems the Internet can change the way you think, or at times make thinking less important. It also impacts your ability to delay gratification and may impact social and empathy development, and there is growing evidence to suggest that today's youth have developed mental health problems from excessive smartphone, screen, and social media use.

However, perhaps the biggest issue with children and teens is the incredible amount of *time* they spend on their devices and the way it impacts their real-time social interactions, physical activity, and relationships. This life imbalance is palpable and can cause real issues in overall health and well-being. This is not to say that everyone who uses Internet and screen technologies needs treatment or even has a problem, but some of the data is certainly a cause for concern.

The goal is to teach children responsible and sustainable use of technology and how to use their screens in limited and balanced ways. See Chapter 15 for more information.

Balancing Technology with Real-Time Living

Anything in excess can produce negative health issues, and the Internet and screen technologies are no exception. Health issues relating to screen use have been well established and may involve reduced and poor sleep, increased stress and elevated cortisol, repetitive motion injury, and neck and back problems. There have been reports of thumb, finger, and upper-back problems from excessive screen use, as well as eye strain and difficulties focusing. Some reports of more serious medical issues include elevated blood pressure, blood clots, weight gain from sedentary behavior, and heart rate issues from dehydration; in extreme cases, people have died.

Health is in large part determined by *balance* in your life; screens and all the content you endlessly consume online may interfere with that balance. But balance is key. It can be hard to maintain because it involves conscious choices, but it is necessary for healthy living. After treating hundreds of patients with Internet and video game use issues and having used the technology myself for the last 30 years, I have come to believe that it is anything but *benign*. It eats up your attention and can rob you of the most important resource in your life — your time — without your even knowing it.

People constantly embrace new technology: faster smartphones (companies are now rolling out 5G), faster processors, bigger and better screens, more apps, and more devices connected to the cloud and running their daily lives. This in some ways represents progress, but in many ways, it is a setback. All this technology requires more and more time and attention to manage, maintain, and learn how to use. I cannot tell you how many times I have had to troubleshoot a problem with some digital device, install a new app, or just change my password for the millionth time. All of this takes time. All of this takes energy. All of this takes attention from other, perhaps more meaningful and satisfying parts of your life. All of this attention to our tech adds up.

Sure, all this convenience is wonderful, but is it really making life happier and satisfying? I'm not so sure. Has all your screen and Internet use really added quality to your life? Just because you can hook everything in your life up to the Internet does not *literally* mean it will improve the quality of your life; you always must ask yourself the question "Will the cost (time) really be worth the ultimate *benefit?*"

Chapter 14 has details on finding balance in screen use. Chapter 16 warns you of how the Internet is only going to become more addictive.

Understanding the End of Privacy as You Knew It

The following sections cover a few of the features of the Internet that many of you may experience as inherent strengths of being online, but many are an illusion when looked at more closely.

Recognizing the myth of anonymity on the Internet

Anonymity is a myth on the Internet as there is no such thing as real privacy online. When we conducted our original research in the late 1990s, several items stood out. Not surprisingly, *perceived anonymity* was a significant factor in contributing to the appeal of the Internet. It seems there was a perception that the Internet and all things communicated online *felt* like they were anonymous. But *nothing* could be further from the truth.

Virtually everything you do and say, type, text, post, search, and scroll online is trackable, traceable, and in some cases recorded and reproducible. That is not to say that everyone is reading and listening to what you do online, but rather to demonstrate that *there is a footprint* of what you do and say online. In general, you should remember that online communications are anything but anonymous. There are countless cases where users are startled to find out that what they do and say online can be accessed by others, and this record appears to be rather permanent. All this is said not to make you paranoid, but rather to be mindful that privacy online is complicated and it's best to remember that.

Nothing online, and I mean *nothing*, is private.

Understanding the disinhibition phenomenon

Another important phenomenon is *disinhibition*. We do not entirely know why people often feel disinhibited or freer to express thoughts and feelings online. Sometimes that can be a good thing, but at times you can say things that might be better left unsaid. When on social media or other communication apps, it's important to remember that the act of communicating online via text seems to bypass some neurological processing that might act as a buffer in verbal interactions; this may be due to differences in processing and less use of the frontal part of the brain that manages impulsivity, although this has not been formally established.

The disinhibition you feel when communicating online can also impact your perception of *intimacy*; we found that when users communicate online, they can feel more connected and intimate with those they communicate with. However, this level of intimacy does not seem to reflect the same types of intimacy that occur through other modes of communication and connection. Although online communication is useful for work and social connection, it can lack some of the cues, boundaries, depth, nuance, and markers that allow real-time connection to establish deeper aspects of relatedness. The ironic fact is that people often feel more intimate more rapidly online, even though it is questionable whether this is analogous to real-time social connection. I have many patients who reported feeling almost instant closeness online only to find it didn't cross over well in person.

Cybersecurity, cyberstalking, and cyberbullying

REMEMBER

Unfortunately, because everything online is vulnerable to hacking, there is a risk that sensitive information such as finances or other personal data can be hacked into, stolen, ransomed, or corrupted. Nothing is fully safe and secure online, and care must be taken to protect sensitive and private information, including, but not limited to, financial data.

The openness and easy accessibility of the Internet, along with the proliferation of social media, have created a firestorm of problems with cyberstalking and cyberbullying, as it is easy to troll someone online and track their digital and social footprint. Even more damaging, however, is the ease and relative frequency with which cyberbullying can occur. When someone says something about someone online, it's difficult to erase the sting and stain of what is said. Perhaps the worst part is that a powerful feature of the Internet is its *broadcast capability* — and this makes cyberstalking and cyberbullying that much easier. Anyone can say anything online, and short of overt hate speech, and sometimes including it, it's easy to get away with saying some psychologically damaging things about someone. This is an all-too-common occurrence for children, tweens, and teens, and there have been numerous cases where significant emotional damage was done and that, in some cases, resulted in suicides.

REMEMBER

The Internet is powerful and can be used to communicate with others, but care must be taken to remember that everything said online is amplified and potentially broadcast to the world.

Chapter **2**

Studying the Biology of Addiction

Addiction can produce numerous negative psychological and behavioral effects. A large part of the addiction process is directly connected to the neurobiology of the brain — the topic of this chapter.

Understanding the Evolutionary Biology of Addiction

The biological factors relating to addiction are primarily associated with the limbic system of the brain. The limbic area is found in the middle of the brain, and it sits on top of the brain stem and lower brain structures (sometimes called the old or reptilian brain) and just under the neocortex (the wrinkled, gray part of the brain — the one you think of when you picture a human brain). *Neo* refers to the newer section of the brain (newer in the sense that it was the last to develop in our evolution).

The limbic part of the brain is much older than the neocortex. Its development is often associated with emotion and more primitive survival functions, and it's sometimes referred to as the mammalian brain. It has a long evolutionary history, having evolved over millions of years in mammals, far predating human evolution. Much of the limbic system's job involves supporting various essential activities that developed to ensure species survival. By comparison, the neocortex is probably only about 200,000 years old, and as such, it's a relatively new arrival on the block when it comes to brain development.

So, what does the limbic area of the brain have to do with addiction? The answer is *plenty,* and it makes a lot of sense when looking at the relative utility of all biological structures and functions throughout the history of our species and understanding that nothing occurs by accident. This all becomes clear when you examine these functions throughout our evolutionary development; the following sections walk you through the brain's evolution when it comes to addiction.

Discovering humans' original addictions

Many brain structures involved in our survival have to do with pleasure, such as the ventral tegmental area, substantia nigra, amygdala, anterior cingulate, prefrontal cortex, hippocampus, and nucleus accumbens (Chapter 3 has more details). However, the nucleus accumbens (NA), located in the brain's limbic system, is the major source of the experience of pleasure.

REMEMBER

The nucleus accumbens is a dense collection of cells that are specifically receptive to dopamine. *Dopamine* is one of the major excitatory neurotransmitters responsible for pleasure and movement. So, why would our survival be linked to an excitatory pleasure neurotransmitter? Well, as nature would have it, a mammal's survival is at least minimally based on the ability to engage in two behaviors on a predictable, consistent, and efficient basis. In other words, nature essentially needs a strong guarantee (from a genetic survival perspective) that can increase the odds of survival. These two behaviors (not surprisingly) are sustenance (eating, and all the behaviors associated with obtaining and consuming nutrition) and procreation (mating). Obviously, adequate nutrition is necessary for effective procreation as well.

The interesting thing here is that food — or more specifically, eating — is highly pleasurable. A strong biological pleasure drive (not just hunger) is linked to consuming food, and this is nature's way of ensuring that food is consumed. So, when we eat, we experience an associated strong elevation of post-synaptic dopamine in the nucleus accumbens; this flood of dopamine is experienced as pleasure. Just think about your last great meal or the intensity of pleasure your dog seems to experience when eating. Food is extremely pleasurable and satisfying to consume for all mammals. Even without language or consciousness to understand this connection, on a strictly biological level, this pleasure increases the likelihood of a behavior being repeated reliably and consistently — thus ensuring survival.

The same applies to procreation and sex. Obviously, it would make sense to make procreation pleasurable on some level. Not just humans, but all mammals have dopamine linked to their mating behavior. This is nature's hedge to increase the odds of fulfilling the biological imperative — to get genes transferred into the next generation. This is the basic survival instinct that accounts for at least part of human behavior — the circle of life, if you will.

Distinguishing the new brain from the old brain

Sometimes we hear references to *new brain* and *old brain*. However, that is not entirely accurate. When we refer to new brain or old brain, what we're really talking about is how the different parts of the brain developed from an evolutionary, biological perspective. As mammals evolved over eons, the brain developed in increasing complexity from the bottom up; this means that as evolution moved forward, it progressed from the lower brain, including the upper brain stem, to the midbrain or limbic system (where the reward centers are), and finally to the latest development in our neurobiological evolution, the neocortex, or new brain.

By the way, it's the prefrontal part of the cortex that helps inhibit the more animalistic survival drives that emerge from the more primitive limbic structures of the brain. And addiction is one of those hijacked drives piggybacked onto the limbic reward center gone awry.

Identifying the Parts of the Brain Involved in Addiction

Many areas of the brain are involved in addiction, and most of them are found in the limbic system. The interesting thing about addiction is that it isn't simply about dopamine and the pleasurable feeling it provides. Numerous other systems are involved with addiction, including the endocrine system and certain hormones, as well as other neurotransmitters. The major areas of the brain involved in addiction include the ventral tegmental area, substantia nigra, amygdala, anterior cingulate, prefrontal cortex, hippocampus, and nucleus accumbens (also discussed in Chapter 3).

The *nucleus accumbens* is the major area in the limbic region of the brain; it's a dense bundle of cells where a very large number of dopamine receptors reside. *Dopamine* is one of the major excitatory neurotransmitters responsible for pleasure and movement, and along with serotonin and norepinephrine, it has a lot to do with mood, reward, motivation, pleasure, and compulsion. (*Neurotransmitters* are chemicals that enable nerve cells to communicate with one another.) The nucleus accumbens is the major traffic center for pleasure for most of the things we do, and as mentioned earlier, the brain is wired to make sure that certain behaviors associated with survival and thriving are pleasurable. It makes sense that nature would do this, as it increases our survival potential. Figure 2-1 shows the reward centers of the brain.

Reward Centers of the Brain

Striatum

Substantia nigra

Frontal cortex

Nucleus accumbens

VTA

Raphe nucleus

Hippocampus

Dopamine pathways
Serotonin pathways

Dopamine Functions
• Reward (motivation)
• Pleasure
• Euphoria
• Motor function (fine)
• Compulsion
• Perseveration

Serotonin Functions
• Mood
• Memory processing
• Sleep
• Cognition

© John Wiley & Sons, Inc.

FIGURE 2-1:
The reward centers of the brain.

REMEMBER

As you read the following sections, keep in mind that addiction involves borrowing those pleasure pathways meant for survival, so in an odd way, addiction is really a natural phenomenon gone awry — much like many medical conditions. It's also important to note that the human brain and nervous system are always about checks and balances — where excitatory neurotransmitters are held in check by other inhibitory neurotransmitters such as GABA (gamma-aminobutyric acid).

Examining the role of neurons and neurotransmitters

REMEMBER

Neurons are essentially the central part of the electrochemical connection process that makes up our nervous system; most of what we're talking about with addiction involves the central nervous system, where neurons do their job. The central nervous system includes the brain and spinal cord. A neuron is made up of three major parts, as shown in Figure 2-2:

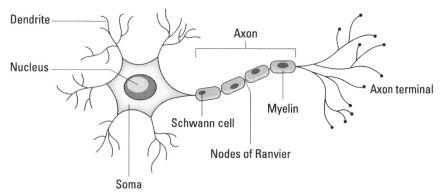

© John Wiley & Sons, Inc.

FIGURE 2-2:
The major parts
of a neuron.

>> Soma, or cell body

>> Dendrites, which are the connecting strands that allow all neurons to interconnect

>> Axons, which are the fibers that act like electrical cables, connecting cell bodies to their dendritic terminals

Obviously, this is an oversimplified explanation, but suffice it to say that the power of the human brain derives not only from its constituent lobes and functional centers, but from how they *all* interconnect and communicate.

Neurotransmitters are the chemicals that allow neurons to communicate and connect with each other. Many neurotransmitters are in the central and peripheral nervous system, the major ones being dopamine, serotonin, norepinephrine, GABA, acetylcholine, glycine, and glutamate. The three that we frequently deal with in psychiatry, psychopharmacology, and addiction medicine are dopamine, serotonin, and norepinephrine, although we have recently begun targeting and integrating some of the other neurotransmitters as well.

The way in which neurons connect is quite amazing and is a combination of electricity and chemistry. Without getting into the weeds here, each neuron is typed to a specific neurotransmitter, and for the most part only uses that neurotransmitter for firing its respective neurons. Here's how it works (see Figure 2-3):

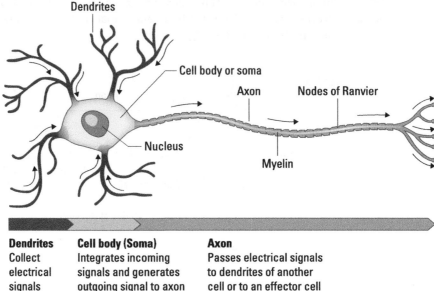

FIGURE 2-3:
A neuron firing.

Dendrites	**Cell body (Soma)**	**Axon**
Collect electrical signals	Integrates incoming signals and generates outgoing signal to axon	Passes electrical signals to dendrites of another cell or to an effector cell

>> An electrical signal moves through an axon (the electrical cable I mention earlier), and this signal travelling along the biological wire is called *propagation.* Some unique aspects of the axon help facilitate this propagation and make it as fast and efficient as possible; a special superfast coating exists along the axon called *myelin,* and this *myelin sheath* allows for a faster electrical transmission.

Nature also uses other tricks to keep nerve signals moving along. Many axons also have built-in speed bumps called *nodes of Ranvier* (shown in Figure 2-2), which essentially allow the electrical signal to jump across the tips of these nodes to speed up transmission. This is in part what allows our nervous system to operate at lighting speed, and this leaping is called *saltatory conduction.*

>> That electrical signal comes barreling down the axon to a gate (actually, a gap) called a *synapse.* When it reaches the synapse (or gap), a seemingly magical thing occurs: The electrical signal is converted to a chemical one by facilitating an *action potential* that commands the *presynaptic membrane* (one end of the

axon) to release its store of the associated neurotransmitter. This transmitter is then taken up (the term is *uptake*) into the post-synaptic receptor and then converted back to an electrical signal; this nerve conduction process occurs millions of times a second. (See Figure 2-4.)

» The leftover neurotransmitter that is not taken up is then either biodegraded (absorbed) or taken back up (*reuptake*) by the pre-synaptic receptor for later use.

Synapse

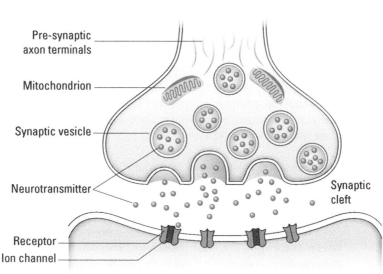

FIGURE 2-4:
Inside a synapse.

TECHNICAL STUFF

Antidepressants, particularly selective serotonin reuptake inhibitors (SSRI), prevent the reuptake of serotonin. This allows more serotonin to be available, which, in theory, can elevate mood and help alleviate depression.

REMEMBER

The king of neurotransmitters related to addiction is dopamine. Although other neurotransmitters are undoubtedly involved with addiction, dopamine constitutes the lion's share of the pleasure drive we see in addiction. Internet screen use elevates dopamine reception at the post-synaptic receptor in the nucleus accumbens, and over time there is an up-regulation of these post-synaptic receptors to manage the increasing level of dopamine present in this area of the brain. See the nearby sidebar "The role of up-regulation and down-regulation in addiction" for details.

THE ROLE OF UP-REGULATION AND DOWN-REGULATION IN ADDICTION

As a person increases dopamine in the limbic reward system of their brain, this translates into increased pleasure; not surprisingly, pleasurable behaviors tend to be repeated, especially when there is an unpredictable or variable element to the pleasurable hit. Biologically, this is where *up-regulation* and *down-regulation* come in.

As the nucleus accumbens is hit with signals that increase dopamine, the post-synaptic receptor binding sites (call them *Pacmen*, as in the video game) first begin to *up-regulate* (become more plentiful) and are constantly occupied by dopamine. The receptor sites then slowly respond less to the same levels of dopamine. This is called *down-regulation* (or *desensitization*), and it makes sense that this would produce *tolerance,* meaning that you would need more stimulation to get the same effect, which I discuss earlier in this chapter. This is because to keep those neurons firing, they now need more stimulation, and this increase produces *down-regulation* or *desensitization.* With technology and screen addiction, this might mean more intense or stimulating content, faster Internet speeds, quicker access, more frequent use, or longer use times — all of which we tend to see with Internet and technology addiction.

All this can lead to the development of *reward deficiency syndrome,* which occurs when the reward center acclimates to that intense level of stimulation and only receives a significant dopamine hit when engaging in excessive or intensive screen time; by comparison, everything else looks and feels flat and uninteresting. This is a very common complaint from parents and family members where their addict loved one no longer takes pleasure in any previously pleasurable behaviors. This is essentially an example of *down-regulation,* in that when those hungry *Pacmen* are consistently fed, they need more stimulation to get the same effect. Real-time living *no longer* provides the stimulation necessary to fire those neurons in the nucleus accumbens in order to help a person feel good or to experience pleasure from normal living. Those receptors become used to being occupied, and they become accustomed to the resulting increased dopamine levels. Unless the person is continuously excited by repeated Internet or screen use, they experience a biological deficiency in the form of psychological symptoms such as boredom and feeling "blah," anxious, irritable, unmotivated, or even depressed.

This is a tough place to be because there is a high likelihood of relapse at this time; we generally want to feel good, and the quickest way to feel good is to engage in the behavior that we know will get that result quickly. It's also a difficult time because the motivation to engage in other real-time behaviors that could eventually prove to be fun and stimulating (dopaminergic) is very low, and it also takes some time for the receptors to *re-regulate.* This creates a challenging period for the addict and their family, as it occurs between the *discontinuation of the old behavior* and its resulting neurochemical effects, and *engagement in the new behavior,* with its more balanced neurochemical state. This new desired state of up-regulation is sort of the opposite of down-regulation on a biological level, and involves getting the dopamine receptors in the nucleus accumbens to *re-sensitize* to more normal levels of *real-life stimulation.* The Internet is very stimulating, and things like video games, YouTube, pornography, or even social media can retrain our nervous system to get used to that stimulation.

Recognizing the Mind and Body Aspects of Addiction

REMEMBER

So, what does all this talk of evolution earlier in this chapter have to do with addiction, anyway? Well, the brain's ability to use dopamine to ensure certain survival behaviors can *also* be stimulated by other pleasurable behaviors. It seems that drugs, alcohol, sex, food, gambling, the Internet, pornography, video gaming, and even social media all elevate our dopamine levels. Addiction is in part related to becoming habitually dependent on this elevation of dopamine; this, along with various psychological, social, and behavioral factors, produces the experience we classify as addiction. What we get addicted to is the *dopamine* and, even more so, the *anticipation* of that dopamine; this, along with other brain chemicals, hormones, and behaviors, essentially hijacks the reward circuits in the brain that are meant to increase our survivability. This means that to the brain, addiction is a misguided piggybacking of survival drives.

This piggybacking of the reward neuropathways leads to a common misunderstanding about the biology of addiction because we often wrongfully assume that we get *addicted* to a *substance.* However, that does not account for behavioral (process) addictions, within which Internet and video game addictions would fall. This is also true for some of the other behavioral addictions, including sex, food, and gambling. The brain, when excited by a video game, doesn't look much different from the brain ingesting cocaine, eating a great meal, or having sex (although there may be differences in the overall level of dopamine innervation). These differences are relevant, of course, and dopamine is involved in all these behaviors and is linked to our survival, as I explain earlier in this chapter.

The following sections go into more detail on how the brain is susceptible to addiction.

Being wired to maximize pleasure and minimize pain

WARNING

The problem here is that in a sense, our brains act as if our addictive behavior is enhancing our health and well-being. However, this couldn't be further from the truth. Perhaps this is partly *why* most addicts take a long time to reach a point where they see how the harm overrides the pleasure they are receiving. Some might call this process denial, where our brain is on autopilot, engaging in a dopaminergic behavior that tricks it into acting on a pattern that maybe hurting us. This is the essence of the addiction paradox: The brain is doing what it knows best, but at the same time, it is creating a significant life problem. This is in part why recovering from an addiction is so hard — why would you want to stop something that feels so pleasurable?

The other factor here is that as you increase activity in the nucleus accumbens, you are simultaneously inhibiting activity in the prefrontal cortex, where reasoning and executive functions are located; the brain is all about pleasure and has less ability (or desire) to apply the brakes once it's in the pleasure cycle.

All organisms are wired to maximize pleasure and minimize pain. This makes sense on many levels, especially when you understand that pleasure equals survival in terms of our brain's evolutionary biology (as I explain earlier in this chapter). Of course, it is understood that we as humans are not simply driven by pleasure; we have many other drives and desires that can transcend our more basic biological nature, but sometimes, some of us get stuck on the pleasure merry-go-round.

REMEMBER

The reward center of the brain is largely selective to the neurotransmitter dopamine, which is an excitatory brain chemical, while the frontal lobes are more inhibitory and are designed to counteract some of these excitatory effects (the main frontal lobe brain chemical is GABA, or gamma-aminobutyric acid). You ideally need both brain chemicals, but addiction can skew your nervous system in the pleasure direction of this delicate balance and tip the scales in favor of dopamine and pleasure, while ignoring the rationality and reason of your frontal lobes. This imbalance is more likely to occur during adolescence and young adulthood due to the brain being immature. The balance between pleasure and self-control is necessary for mature development, but too much of a good thing (dopamine pleasure) can increase the likelihood of an addiction and produce a negative impact on real-time living. Flip to Chapter 4 for more details.

REMEMBER

Addiction is not simply the physical dependence on an intoxicating drug or behavior. Rather, addiction involves changes in motivation, learning, mood, life balance, social relationships, negative health impacts, and decreased work or school productivity. With specific regard to Internet addiction, there can be an all-encompassing preoccupation with screens and the content consumed online.

OUT IN THE WILD: ADDICTIONS IN ANIMALS

Addiction appears in nature in some ways, but not exactly as we see it in humans. There are many reports of animals that seek out psychoactive plants and herbs that essentially get them high; however, keep in mind that addiction is not simply about a *substance* or a *behavior*, but rather about an *imbalance in life functioning that involves mind, body, and behavior*. The evidence suggests that animals do seek out plants that are intoxicating, and the effects seem obvious — for example, elephants lying about lazily after consuming an intoxicating plant — but I am not aware of any evidence that engaging in these behaviors becomes as problematic as it can be for humans.

The reason for this difference seems clear: Humans require a far more complex set of behaviors to function than do animals, and we require balance across many life domains to stay healthy.

Exploring tolerance, withdrawal, triggers, and relapse

Addiction is often associated with several key behaviors that frequently highlight the addictive process. Some of these processes are relevant to Internet addiction, but in a manner that is specific to screen overuse.

Tolerance and withdrawal

Tolerance and *withdrawal* are terms that we use in addiction medicine to describe the impact of a substance or behavior on our physical and psychological functioning:

>> *Tolerance* is a technical term to describe how the body (including the brain) develops decreased sensitivity to the intoxicating effect of a substance or behavior (such as Internet use); the effect of this decreased sensitivity is that it takes more of a drug or behavior to achieve the same degree of intoxicating effect. By intoxicating, I include the subsequent elevation of dopamine. This is a natural process and involves regulation of post-synaptic receptors mostly found in the nucleus accumbens. It's important to note that most of the time, we are unaware that we're developing a tolerance; to others who may be observing, it may seem obvious, but this is often a gradual process that goes on without the addict being conscious of it.

>> *Withdrawal* involves the physical and psychological reaction to decreasing or discontinuing a substance or behavior; as that dose decreases or stops, certain symptoms are experienced. The body and mind have acclimated to a certain level of intoxication and pleasure. Regarding Internet and technology use, the withdrawal effects might be expressed by reduced frustration tolerance, increased irritability, anger, anxiety, social isolation, and depression. Physical expressions of anger and rage have sometimes been noted, including breaking furniture or punching holes in walls. I have seen these withdrawal effects in patients over the years, and they can be quite disheartening to both patients and their families but are typically short-lived.

In some cases, especially in adolescents and young adults, withdrawal effects can become quite concerning. Obviously, *safety is the number one goal here,* and if your child is threatening to self-harm or is damaging property, it might be necessary to obtain support by calling your local crisis intervention line or the police, or by going to the nearest emergency room for an evaluation. Obviously, this is not a desirable place to be, but I have seen some younger patients become so angry and in acute withdrawal that they lose their composure and self-control, and in those cases, it is always better to be safe and to protect yourself and your loved one.

The degree and amount of tolerance or withdrawal *are not* the determining factors for evaluating the degree or severity of an addiction. The amount of withdrawal *does not* indicate whether there is a diagnosis of Internet addiction. Addiction is a complex, *biopsychosocial* state that involves disruption in physical health, mood, motivation, time management, social relationships, and productivity. You can have an addiction to the Internet and have little or no withdrawal. Indeed, if getting through withdrawal or detoxing from the Internet, video games, or other screen behavior were sufficient, then no one who discontinued Internet or video game overuse would be experience addicted behavior again or experience a relapse, but this isn't the case.

Triggers and relapse

Triggers are part of every addiction. A *trigger* is a psychological tickle that excites the reward system in your brain, which remembers the pleasure it felt when it engaged in the addictive behavior. I say "remembers" because part of the reward system includes the hippocampus, which is the part of the brain's limbic system that is responsible for memory. The other interesting thing to note is that a *trigger* can elicit an *anticipation response* for getting that dopamine hit, and in many cases, that *anticipatory dopamine hit* can be *twice* as innervating as the dopamine hit from the actual behavior. A trigger is a powerful reminder of a past pleasure and can lead to a complete relapse into an addiction pattern.

Relapse is straightforward — it involves the reactivation of an addiction pattern. Sometimes relapse occurs for no obvious reason, and sometimes it occurs because of a clearly identified trigger. Triggers can be reminders of the addiction or pleasure found in the behavior, or they can be an emotion or a physical need of some type, including being tired, sad, lonely, or hungry. Basically, any imbalance in our well-being can be an indirect trigger for relapse. A big trigger for Internet addicts can simply be boredom, or simply seeing their screen device.

TIP

Some of the most common triggers can be memorized using the acronym HALTS: *Hungry, Angry, Lonely, Tired,* or *Stressed.* I would add that there are many other potential triggers as well, which often include proximity to our screen device, reacting to a time of day, or having something frustrating happen to us.

REMEMBER

It's important to note that relapse *does not mean failure.* Addiction often includes many attempts at recovery before success can be achieved; with Internet and technology addiction, there is an added complexity in that one *must define* what constitutes a relapse. Because it would be difficult to avoid screen use completely, a different definition of relapse must be defined for each person; typically, this includes deciding on a reasonable and healthy amount of total screen use, including a time limit or avoidance of specific content areas such social media, video gaming, pornography, YouTube, or other highly compelling websites or apps. The goal here is mediated and sustainable Internet and screen use, not abstinence of all screens, because that is impractical.

Comparing physiological dependence versus addiction

Physiological dependence involves the body (and brain) becoming used to a substance or a behavior. It is most typically associated with drugs and alcohol, and withdrawal is often medically treatable as a temporary condition that requires monitoring and possible pharmacological intervention. With certain types of physiological withdrawal, such as with alcohol, care must be taken to ensure against medical problems. With process and behavioral addiction such as Internet and technology addiction, some physiological withdrawal still occurs, but medical supervision and intervention are typically not needed. That does not mean professional guidance from a mental health or addiction professional is not needed — this may be necessary, and sometimes medications may be utilized to assist in both the withdrawal and recovery processes (see Chapters 11, 12, and 13 for more on diagnosis, self-help, and treatment).

Although there are notable withdrawal effects with Internet and technology addiction, these are typically short-lived (a matter of weeks). It is hard to separate physiological dependence from psychological or behavioral dependence, as the mind and body are not actually separate, but rather are an integrated system — addiction always affects both. However, physiological dependence often has to do with other organ systems of the body besides the central nervous system, and it accounts for the immediate physical withdrawal symptoms that we see with drugs and alcohol for a period after discontinuing a substance (symptoms that may require medical intervention).

REMEMBER

Physiological dependence is *not* addiction; addiction is always a more complex, biopsychosocial process that involves numerous aspects of physical, behavioral, and psychological functioning.

Looking at addiction as a mind and body phenomenon

REMEMBER

Addiction is always a mind-body phenomenon and typically involves numerous aspects of our functioning. You cannot have an addiction without impacting life functioning in some way. Obviously, addictions vary in intensity and life impact, and so there are differences regarding the need for, and type of, treatments. The main thing to remember is that addiction always creates some consequence for how one *lives* and functions in their life, and always produces limitations, compensations, and challenges that are both physiological and emotional. We are often unaware of how impactful our addictions can be, and Internet addiction is no exception. Most of our patients are in some degree of denial regarding how impactful the addiction is for them and how it has limited and affected their life, which, as noted earlier, can be frustrating to loved ones.

Discovering how addiction is a normal medical problem of living

REMEMBER

As mentioned earlier in this chapter, addiction is really a part of normal brain reaction and functioning and is hardwired into our brain's development. We're designed to engage in pleasurable behaviors that were originally linked to enhance our survival, such as food and sex. Nature created the reward system in our brains (and in most mammalian brains) to facilitate our engaging in survival behaviors that are linked to pleasurable dopamine. These ancient parts of our brains are what make us all susceptible to addiction. The old stereotype that you become addicted only if you have an "addictive personality" is not based on sound science; the fact is that if you have a brain, you have some susceptibility to addiction.

Granted, there are genetic, epigenetic, and environmental contributing factors to addiction, but as I've said, the basic potential for addiction is hardwired into all of us. There is always an evolutionary part to addiction, where modern pleasurable triggers such as drugs, alcohol, calorie-dense foods, pornography, the Internet, video games, social media, shopping, and gambling all piggyback on those original dopamine survival pathways.

Addiction is never solely about willpower, character, motivation, or honesty. Society often labels an addict as having deficiencies in these areas because the addict may appear to be displaying socially undesirable behavior. Addiction is a neurobiological disorder with significant psychological, social, spiritual, and physical impacts; no one chooses to be an addict to any substance or behavior (although they do have to choose not to be one). An understandable error is often made by those who love an addict. I see parents and family members become angry at their loved one for being an addict and engaging in negative behaviors that are typical of addicts. For instance, addicts almost always lie. Mainly, they lie to themselves in the form of rationalization and denial. This is necessary to continue their addictive behavior. After all, how can you be an addict and at the same time be honest with yourself that you are suffering from significant life consequences from watching 10 to 14 hours of YouTube, playing video games all day and night, never putting your smartphone down, or living on social media? It is hard to objectively see the deterioration in your relationships, your school or work performance, or your health, sleep, and personal hygiene.

REMEMBER

Addiction is a normal human medical problem. And, as with all medical problems, people may need help. It is important not to judge our addicted loved ones, although it is understandable how they become frustrating to deal with. Illness is unfortunately part of the normal human condition, and addiction is no exception.

Chapter **3**

Understanding Why Kids Are So Susceptible to Internet Addiction

A nyone can develop an addiction at any stage of their life, but children and adolescents seem to be more susceptible to addiction. This increased susceptibility should be no surprise, after all, as these are formative years — when young people acquire the habits they will carry for the rest of their lives. The brain, where addictions are formed, is more malleable, more impressionable, during these early stages of development; in addition, the areas of the brain that are responsible for executive skills and judgment are not fully developed until around age 25. The inexperience of adolescence compounds this vulnerability, and this is also the same period of life when children and teens are introduced to the various objects of addiction — alcohol, nicotine, drugs, sex, video games, Internet, smartphones, social media, pornography, and so on.

Understanding the process of addictive behaviors is valuable in both preventing and overcoming addictions of all kinds, especially in the context of children and adolescents. When you have a clearer understanding of what is going on inside the mind and body of your child, teen, or young adult as they interact with the world, you're better equipped to be empathetic to their addiction, and to support interventions to effect positive change.

In this chapter, I reveal what's going on inside children and adolescents during this time when they're most susceptible to addiction, and I look at how certain internal and external factors may contribute to the problem.

Exploring the Biological Basis for Addiction in Children and Teens

All addictions are centered in the brain and are reflected in our behavior in the form of habits, so the brain is the logical place to begin the investigation into why children and teens are particularly susceptible to addiction.

REMEMBER

Simply put, what happens in the developing brain to increase susceptibility to addiction is that the pleasure and reward centers (in the limbic system) begin to fire heavily *before* the brain's mechanisms to buffer that desire for pleasure are fully formed. In other words, desire and the capacity to feel pleasure (through increased dopamine receptors in the nucleus accumbens) are off the charts during adolescence — and at the same time the neurological capacity for good judgment and self-control (found in the frontal lobes) is just starting to develop. It's like hitting the gas pedal but without a set of brakes! When the desire for pleasure exceeds the capacity for self-control, you have the perfect environment for developing addictions.

In the following sections, I take a deeper dive into the pleasure centers and control centers of the brain to provide a clearer picture of the biological factors that can drive the development of addiction in children and teens. I also explain the role that sex hormones play in compounding the problem.

Getting to know the brain's pleasure centers

Several areas of the brain are involved in addiction, including the ventral tegmental area (VTA), substantia nigra, amygdala, anterior cingulate, prefrontal cortex, hippocampus, and nucleus accumbens (see Chapter 2 for details on the brain). The *nucleus accumbens* is the main area in the limbic pleasure region of the brain; it's a dense bundle of cells where a very large number of dopamine receptors reside. *Dopamine* is one of the major excitatory neurotransmitters responsible for pleasure and movement, and along with serotonin and norepinephrine, also has a lot to do with mood, reward, motivation, pleasure, and compulsion — as well as numerous frontal lobe functions. (*Neurotransmitters* are chemicals that enable nerve cells to communicate with one another.)

REMEMBER

The development of the nucleus accumbens and the dopamine reward system seems to mature in its dopaminergic efficacy sometime in the mid-teens, resulting in a powerful desire for fun and pleasure.

Buffering the desire for pleasure

To simplify, the brain has two primary mechanisms for buffering the desire and drive to experience pleasure: the frontal executive areas of the brain, and the inhibitory neurotransmitter gamma aminobutyric acid (GABA). Unfortunately, when activity in the brain's pleasure centers begins to peak in the mid-teens, neither of these inhibitory mechanisms is fully developed.

The frontal lobes

REMEMBER

The frontal lobes (shown in Figure 3-1) start to develop and mature during the teen years but aren't fully formed until around the mid-twenties (give or take). This is, in part, why the teen years can be rocky at times. Desire for pleasure exceeds the judgment needed to control it, often resulting in what might be considered *irresponsible decisions* and *behaviors.* This seesawing of pleasure and control is further complicated by inexperience, impulsivity, and feelings of invulnerability that permeate adolescence. The timing of when the pleasure drive peaks and the frontal lobes fully develop varies among individuals, but this process is part of normal biopsychosocial development.

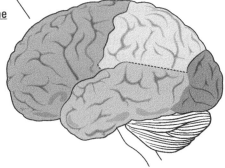

Frontal lobes

Executive Functioning - Dopamine

- Planning
- Problem Solving
- Motivation
- Judgment
- Decision Making
- Impulse Control
- Social Behavior
- Personality
- Memory
- Learning
- Reward
- Attention

FIGURE 3-1: The frontal lobes, executive function, and dopamine (DA).

© *John Wiley & Sons, Inc.*

THE EVOLUTIONARY POWER OF ADDICTIONS

Many theories have been proposed to explain reduced access to the frontal lobes when people are engaged in addictive behaviors, but it's likely due in part to the evolutionary and biological origins of addiction. The powerful dopamine pleasure center in the brain evolved because it helped ensure survival of the species. Think about it; food and sex are associated with some of the biggest releases of dopamine in the body, and they are two factors that are essential to human survival.

That food and sex become addictions for many people is no surprise. In fact, if you look at the overall negative health impact, food addiction alone likely dwarfs all other addictions combined.

Eating and sex are pleasurable because both are *essential* for the survival. They're pleasurable, and hence dopaminergic, because our brains want to make sure we engage in these activities, and what better way to make sure we do something than to make it incredibly desirable?

It may be that when these powerful pleasure centers are activated, innervation of frontal control circuits that provide reasoning, judgment, and executive control is temporarily reduced; it is my theory that this occurs to facilitate survival behavior powered by the pleasure center in the nucleus accumbens. Why would this occur? It makes sense, from an evolutionary perspective, that thinking and reasoning behaviors (which are slower) could limit dopaminergic survival behavior; this would be contrary to engaging in the immediate-acting survival behaviors of sex and food. If you are hunting or finding food or attempting to have sex, anything that could slow this process down would reduce the likelihood of survival. It so happens that the same areas of the brain that makes survival behaviors pleasurable are involved with addiction.

Unfortunately, whenever anyone engages in addictive behavior, the body tends to diminish and limit access to the frontal lobes of the brain. Disengaged, the frontal lobes can't buffer those strong desires for pleasure that are often characteristic of an addicted state. Why the body shuts down access to the frontal lobes when a person is engaged in addictive behavior is not fully understood, but one possible explanation is presented in the nearby sidebar "The evolutionary power of addictions."

GABA

Another reason the brain does a poor job of buffering the desire for pleasurable activities in children and teens is that the level of the inhibitory neurotransmitter GABA is low during this period. As I mention earlier in this chapter, dopamine is like the accelerator of the pleasure drive, and GABA is like the brake. The body uses dopamine and GABA together, like an adaptive cruise control, to modulate pleasure drives. Children, adolescents, and young adults have much more dopamine compared to GABA and therefore are more prone to pleasure-seeking (and hence, addictive) behaviors. Various electronic amusements (Internet screens) can provide the means for satisfying this craving for pleasure.

TECHNICAL STUFF

One of the interesting things researchers have noticed about adolescence and addictions is that the prevalence of some substance-based abuse and addictions has declined over the last decade while other indications of reduced mental health have increased. One of the theories noted by Dr. Nora Volkow, from the National Institute on Drug Abuse (NIDA), is that perhaps all the exciting innervation of dopamine from screen use has somehow served to substitute the role those various substances might previously have fulfilled. So, although that might be a good thing, there are other mental health markers that have increased, such as depression, during the same period. These observations are correlational, but they pose some interesting ideas regarding the drug-like effects that screen use can have.

Tossing sex hormones on the fire

During puberty, the period during which adolescents reach sexual maturity, the proliferation of sex hormones, particularly estrogen and testosterone, accelerates. In males especially, the increase in testosterone increases desire and the potential for taking risks and engaging in pleasurable and stimulating behaviors, all of which correlate with the development of the dopamine pleasure/reward circuits described earlier in this chapter.

REMEMBER

Several factors are now at work ramping up the desire and capacity for feeling pleasure, while the body's buffering mechanisms are still in development. The convergence of accelerated development of the brain's pleasure centers (increased dopamine), underdeveloped frontal lobes, and immature gabaminergic functions, along with increased sex hormones, creates a perfect storm for abuse and addiction in adolescents and teens, which extends through the young adult years.

Seeing the Impact of the Internet and the "Maybe" Factor on the Developing Brain

A susceptibility to addiction doesn't necessarily result in addiction; there are many factors, including availability, ease of access, and emotions. Something must *trigger* it — an exposure to something desirable and pleasurable, along with other factors. The use of something pleasurable alone also doesn't necessarily produce an addiction; most of us can engage in pleasurable activities and behaviors without becoming addicted. Rather, addiction is likely an interplay between opportunity, environment, genetic predisposition, biology, and various psychological factors — a perfect storm of sorts.

In addition, our brains learn to anticipate a pleasurable experience and to expect another fun time, and it seems this expectation may elevate dopamine even *higher* than the actual behavior itself. When you engage in a pleasurable activity, such as playing a video game, interacting on social media, watching a video, surfing on the web, or scrolling on your smartphone to satisfy your curiosity or manage your boredom, you're elevating the release of the pleasure chemical dopamine in your brain. If you find something very pleasurable or interesting while you engage in those activities, then your dopamine level may become even more elevated. You may also be shutting down the connection to your brain's frontal lobes and suppressing the release of GABA, the brain's inhibition factor. (See the nearby sidebar "The evolutionary power of addictions." Dopamine, the frontal lobes, and GABA are all covered earlier in this chapter.)

REMEMBER

A pleasurable experience alone is probably not enough to trigger an addiction; if you experience a pleasurable behavior consistently, it is likely that eventually, you will become desensitized (bored) by it. Addiction is in part created by *intermittent* and *variable* pleasurable reinforcement, or what I like to call the *maybe* factor, and the brain's pleasure center loves *maybe* like nobody's business. When you or a loved one goes online through any device, it's like sitting down in front of a slot machine and spinning those reels for the *possibility* of hitting it big. Every time you see something you enjoy, whether it's a *like* on social media, a comment, a news feed, a photo, or a video, you are getting some small hit of dopamine. Notice that I said *possibility*, because if it were a certainty, your brain would get bored and come to expect the reward. When winning is a sure thing, the game soon loses its addictive impact.

The power of *maybe* to light up the pleasure centers of the brain is also observable in the pursuit of likes, comments, or follows on social media. Think of the drive people to get *liked* on Instagram, Facebook, or TikTok. Say you posted an amazing photo or video of you petting a baby kangaroo in Australia — the animal is taking the food right out of your hand! You love the photo, and you think others will, too. You're excited while you wait to see how many likes you'll get (or if you'll get any). All this excitement produces a small neurochemical rush, and it's all due to the power of *maybe*. If you knew for a fact that you would get a certain number of likes or comments in a certain time frame, you probably wouldn't post — because it was expected, and less innervating.

WARNING

The Internet, and especially social media, can be a search for our own echoes. A problem with waiting for likes and comments is that there is an unfortunate tendency to base your self-worth on the *reflected echoes* of how people respond to what you post. This pursuit of the *like* drives further posting and scrolling to a point where it can become addictive. Waiting for social media likes, comments, and follows can reinforce basing your self-esteem on what others think, and not on your own experience.

Recognizing Lack of Experience as a Contributing Factor to Addiction

REMEMBER

Addiction initially thrives on inexperience. Through experience, people acquire the skills to maintain balance between purely pleasurable activities and unpleasant activities (delays of gratification) that may be helpful for future success. They learn through the consequences of excess and through negative consequences that result from lost time and life imbalance. Addiction, in a sense, is a form of *incomplete learning*, as you're only experiencing the pleasure and often ignoring the negative consequences of your actions and behaviors. Through mindful moderation, experience, and conscious thought (which requires an unintoxicated brain), you can learn about your addiction and how to change it; it's also important for healthy development that children and adolescents experience the world and its natural consequences and real-time rewards.

Watching out for overprotection

Out of love and concern for your kids, you may think that it's a good idea to protect them from all of life's ups and downs. This is certainly understandable, but is it always best for your child? It seems that parenting has become safety-focused, often to the point of overprotecting children and robbing them of the opportunity to experience mistakes and their consequences. While, of course, you don't want anything dangerous or damaging to happen to your children, there is a point where such protectiveness does more damage than good and potentially blocks them from learning how to manage the ebbs and flows of life. This can also limit your child's willingness and ability to take the necessary risks required to reach maturity and launch into adulthood.

Through their experiences, children encounter limits, consequences, emotions, socialization, compromise, and how to manage their minds and bodies. They come head-to-head with their fears and develop the resilience and courage they need to face challenges. As children learn, they begin establishing a balance between the strong limbic (dopaminergic) desires for pleasure, and the gabaminergic inhibition of the cerebral cortex and frontal lobes (all covered earlier in this chapter). As their brains' executive functions strengthen and become better equipped to counteract desire for excess pleasure, children can engage in more rational internal discourse, such as "Yes, that would be a lot of fun, but no, I'm not going to stay up all night gaming because I have an exam to study for and I have to be up early." Or they will become able to bargain with themselves with a delayed reward after accomplishing a necessary (but less pleasurable) task.

Thwarting normal development

WARNING

Internet addiction (and all addictions, for that matter) is a form of escapism, which thwarts the ability to learn in healthy ways. Addictions temporarily thwart the opportunity for real-life experiences and to learn from the natural consequences (positive, negative, and neutral) of various choices and behaviors, and overindulgence in anything that separates a person from balanced real-time living can hamper personal growth and development. It's important to note that the Internet and screens themselves are not the issue here; it's the powerful, mood-altering, and addictive pleasure that can skew healthy choices.

Most addictions tend to thwart normal development. Addiction has a way of side-tracking typical developmental milestones, especially around social, occupational, and academic pursuits. In the clinic where I work, it's not uncommon to see patients who seem several years their junior with regard to developmental milestones and tasks, relative to their chronological age. They will have delays in normal things such as getting a driver's license, dating, applying to college, or getting a job. One of the interesting changes I've seen from heavy Internet and screen use is the delay or avoidance of getting a driver's license; one of the theories for this (which is more common than you might think) is there is *less desire to go anywhere because everywhere comes to them online!* Please keep in mind that even if such delays occur, it does not mean they won't be able to get back on track, as many of our patients do — most addictions throw people off their normal developmental line, but many *can* and *do* get back on track.

At the residential and outpatient treatment centers where I work, I see the same scenario repeatedly: The patients do not, or cannot, master the challenges of growing up and becoming autonomous, self-regulating adults — in short, they have trouble launching into adulthood. Internet screen use can block the adolescent or young adult from mastering basic life skills; it creates developmental and neurobiological detours that impede growing up and limits their ability to experience the pleasures and responsibilities of *adulting.*

Addiction and delayed development create a vicious cycle. Impairment in frontal lobe activation impairs a person's ability to perceive the *risks* or *consequences* of an addictive behavior, which, in turn, reduces their *inhibitions* to continue to engage in the addictive behavior. The person can then develop a heightened perception of invulnerability and a reduced appreciation for the consequences of their addictive behavior. This lack of appreciation of their behavior, and subsequent consequences, contribute further to the addiction cycle. This is an example of how denial and a lack of objectivity are a significant part of addiction.

WARNING

Please recognize that much of this process is unconscious and largely neurobiological, and that often, normal stages of development, life stressors, trauma, and other medical or psychiatric issues can contribute to the potential for addiction. Many parents and loved ones assume that their child's excessive or addictive use of technology is *volitional* when, in fact, it is *not;* although it may start as such, it can quickly degrade into a deeply addictive pattern of behavior with little self–awareness, often accompanied by denial. Parents often adhere to the belief that video games and screen use are under the *conscious control* of their child when nothing could be further from the truth. Their anger at their kids can be amplified by an erroneous assumption that their children should be able to *control* their screen use, when in many cases they can't. This process can devolve into frequent struggles over technology, including fighting, hiding modems and devices, and inconsistent attempts to limit a child's screen use. These techniques typically fail as there needs to be a more *organized* and *consistent* treatment plan in place to effect lasting change.

THE MYTH OF THE INTERNET AS A POSITIVE TOOL FOR ACADEMICS AND PRODUCTIVITY

One challenge in the battle against Internet and screen addiction is that the Internet is often promoted as a tool for improving productivity and academic performance in school, but this is not actually the case. The reality is that most of the time children and adolescents spend on their devices is not necessarily *productive* nor *academic.* Furthermore, the relationship between academic performance and screen use is far from conclusive. Increased distractibility, time distortion, reduced attention span, boredom intolerance, and a variety of cognitive and psychological changes seem to be related to higher levels of Internet screen use. Additionally, on average, school-age and college-age kids typically use their screens many hours per day, with only a small percentage of that time devoted to academic pursuits.

Obviously, during the COVID pandemic, use of screens became a virtual lifeline for school and social connection. This increased reliance on screens has made balanced use even more complex, but it doesn't change the fact that there are many ways to live with less screen time, including spending less time on smartphones.

The Internet and all the screens connected to it are not inherently problematic, but rather serve to be a potent distractor from less stimulating activities, such as school, sleep, self-care, work, chores, and so on, and there is a tendency to find previously motivating pursuits less rewarding due to *reward deficiency syndrome.* This topic is discussed in more detail in Chapters 2, 4, and 10.

In addition, while computers and the Internet are often presented as a necessary tool for learning, you rarely hear about them being used as a popular tool for cheating, even though it is a common practice. As reported by Common Sense Media in 2019, teens are using their smartphones to "cheat" on homework, tests, or other assignments.

Exploring Other Factors That Contribute to Addiction in Children and Teens

The interaction between the brain and the Internet is only one aspect of developing an addiction to screen technologies. Numerous other factors may also contribute, such as those in the following sections. (A major factor — children's and teens' lack of experience and natural consequences — is covered earlier in this chapter.)

The constant presence of smartphones

Screens connected to the Internet (especially smartphones) are the norm for how youth socialize, interact, and spend *significant amounts* of their discretionary time. On average, teenagers spend five to six hours a day using social media, Snapchatting, Instagramming, texting, and scrolling on their smartphones — frequently sacrificing sleep, academics, and physical activity to do so. Smartphones and other screens have risen from being tools of productivity and utility to being a *peer-based* requirement for living. They are no longer simply devices, but have become a required, stimulating, and addictive possession. Indeed, you would be hard-pressed to find a teen (or tween) without a smartphone in the United States.

According to Statistica in 2020, about 100 million children and younger adults have smartphones, with nearly 6 million under age 11 having one. Another important point, according to the Pew Research Center, is that about 56 percent of parents feel that they themselves are using their smartphone too much, and about 36 percent feel they are on social media too much. These are adults we are talking about who have fully developed brains, and they feel they are not able to control their smartphone or social media use. This gives you some perspective on how potent these technologies really are; considering this data, it makes sense that our children, tweens, teens, and young adults have a guarded prognosis in being able to manage these technologies effectively.

Smartphones have become more like clothing accessories than communication devices, and it's only a matter of time before people are literally wearing them as opposed to holding them. Because a smartphone is seen as more than a device, it's important to have the newest and best options (even though they essentially do the same things). There is often excitement around apps and phone capabilities offered on new models that are quite expensive, and the smartphone and cellular industry spends millions on marketing, promoting the need to have the latest and greatest device.

As I mention earlier in this chapter, the good news is that your kids may be less likely to want you to buy them a car, because data show that, on average, teens are obtaining their driver's licenses much later than previous generations. Why go anywhere when anywhere comes to them?

The "need" for technology

The developers of the Internet viewed the web as a utilitarian tool. They never envisioned it as entertainment or a detour from life. The Internet has now almost morphed into having an identity of its own. It's no longer a tool used to accomplish something, but rather it's equivalent to what a car used to be to a teen or young adult. I've heard others refer to the Internet as essentially a virtual mall, and kids now hang out in cyberspace instead of physically. Some children likely feel more at home on their device than with people. Talking on the phone is done. Conversation has changed. It has been replaced by chatting, texting, IMing, DMing, snapping, commenting, liking, following, and Instagramming.

WARNING

When teens use the word "talk," they don't mean verbal communication; rather, they mean texting, chatting, or snapping. It's likely that your child's ability to communicate verbally is either underdeveloped or has atrophied to the point where they may need some skills development. Words have been abbreviated to the point where a few letters or symbols (emoji) are used instead of complete words, sentences, and thoughts. Everything is streamlined and cut short. Although this may or not be equivalent to more in-depth forms of communication, it may be this generation's form of social and cultural differentiation. Screens have become integral to personal expression, and they seem to be *augmentations* to help express aspects of their identity — such as personal appearance, fashion, politics, friends, or interests.

The growth of social alienation

WARNING

Growing levels of social alienation, seemingly fed by overuse of Internet technologies, place children, tweens, teens, and young adults at an increased risk of developing addictions and other mental health issues. Recent data show markedly greater levels of psychiatric issues, such as depression, suicidality, identity concerns, and social alienation, that not surprisingly correlate with the rapid adoption of the smartphone over the last 13 years. Social media use appears correlated with reduced social empathy and has increased and promulgated cyberbullying and trolling to concerning degrees.

Video games and screen distractions may often fill a void and provide a means of engagement and feelings of acceptance when the user lacks the self-confidence or social skills comfort. Internet activities are easy ways to avoid boredom and unpleasant tasks, and to connect with peers. This is not all bad, and many good things can and do come from this process. Excelling on a video game affords an enhanced level of accomplishment, which a parent cannot ignore; often, your child feels a sense of mastery and success in their gaming, relative to other aspects of their life. You may hear your child tell you how *good* they are at the game and how highly they are ranked or appreciated by fellow gamers. As a parent, it's important for you to understand and acknowledge how much their *sense of self* may be enhanced by gaming.

REMEMBER

The goal is to help expand your child's level of competence and self-efficacy beyond the game or their screen activities. Initially, this may seem like a tug of war, but ultimately, it's a necessary part of the recovery and growth process. Part 3 has more information on diagnosing and treating Internet addiction.

Technology as a coping strategy

WARNING

If children or teens suffer from physical, mental, or emotional trauma, they may turn to addictive substances or behaviors as coping mechanisms — especially if they haven't yet developed other coping strategies. Addiction to anything almost always starts as self-medication (or avoidance) to manage uncomfortable feelings or circumstances; ultimately, it starts out as a *solution* to a problem but then becomes a *problem* unto itself. Again, the Internet is not inherently the problem, but it's a very powerful activity that can become a problem.

Genetic influences

Children of addicts may be more susceptible to developing addictions, and the addiction of choice may differ from that of the parent. For example, if the parent has an addiction to alcohol, the child or teen may develop an addiction to video games, gambling, or another behavior. The data on genetic influences on addiction seem to show some inheritability for general addiction, but even so, this only accounts for some of the reasons why addictions occur. The best thinking seems to point to addiction being a multi-causal problem with many variables involved and Internet and video game addiction is likely to be no different.

An intolerance of boredom

The growing intolerance of boredom is a significant factor for Internet addiction. While screen time may not always provide anything of great value, it can fill a void and provide an escape from the *uninteresting* moments in life. Many children, teens, tweens, and young adults are unable to sit for even a moment without reaching for their phone (this is probably true for many adults as well). When was the last time you waited in line at the post office or in a waiting room and didn't pull out your phone? The ability to tolerate time without distraction has seriously eroded.

You may have noticed, in both yourself and your children, a growing intolerance of even a moment of doing nothing, and the easiest thing to do to escape that discomfort is to pick up your phone. However, these are moments that provide a potential opportunity to experience the spark of self-reflection or creativity, or the impetus for social connection. In short, not giving in to the urge to pick up the phone creates new neuropathways that give your child or teen an opportunity to sit with themselves for a moment and to be with themselves on a deeper level.

If your child's attention span seems shorter than yours was as a child, this is not your imagination, as intolerance of boredom and impatience has become a significant side effect of screen and technology use. I've seen patients who can't tolerate even two seconds without picking up their phone; their attention often seems to be elsewhere, and their patience and ability to delay gratification appear to be diminished. The net effect (pun again) of this pattern of endless screen use further serves to solidify addictive patterns and to weaken more positive self-management strategies.

REMEMBER

Boredom is necessary for achieving life balance. Screens can rob youth of the ability to tolerate the discomfort of boredom with the promise of something shinier and more attractive — but this occurs at a cost of endless distraction.

Chapter **4**

Discovering What Makes the Internet and Smartphones So Addictive

The Internet is no longer new. When I started researching and treating Internet addiction, it was the mid- to late 1990s and the Internet was still the digital Wild West. Twenty-five years ago, few understood that this new, miraculous modality of communication and commerce could be addictive, but we did understand that it was very appealing.

My clinical and academic medical background is in addiction medicine; I was trained in the diagnosis and treatment of addiction, and I've also developed an expertise in sexual medicine and human sexuality — including compulsive sexual behavior. As the Internet burst across the globe, I began seeing patients who appeared to be having some difficulty with various aspects of their Internet use. Because the Internet was so new (and Wall Street was in love with its financial promise), most people did not see that all was not perfect in cyberspace.

In the early to mid-1990s, we were still using dial-up Internet that worked at a snail's pace, but I still recall getting my first decent computer hooked up to the Internet and the welcome allure of those *beeps* and *buzzes* it made as our modem connected to the World Wide Web. I felt as if I were taking a trip somewhere and connecting to faraway people and places. It was exciting. Back then, online shopping and commerce were barely in their infancy; these were pre-Amazon days (if you can imagine that), and there wasn't a lot happening online back then. But even before social media, Wi-Fi, smartphones, and high-speed Internet, you could feel how captivating this technology was.

In this chapter, you discover the common features and factors that make this technology so addictive. Although this problem is now largely accepted in the medical field, I want to offer what I've learned over the last 25 years of researching and treating it.

Eyes on the Prize: Factors Involving Focus on a Screen

Why do we like looking at our screens so much? It may seem intuitively obvious why video gaming is addictive. After all, it is stimulating, endlessly variable, rewarding (the fancy medical term is *dopaminergic*), easily accessible — and fun. But in both my research and clinical practice, I find many areas of Internet screen use to be addictive as well. These include video platforms like YouTube, social media, video gaming, Reddit, online shopping, online gambling, information scrolling and surfing (kind of equivalent to TV channel surfing from the old days), and, of course, pornography. This section looks at the main reasons why people can't help focusing on their screens.

Examining ease of access and near-constant availability

Ease of access has always been an enabling factor in addiction. In substance-based addictions, the availability of certain drugs or alcohol could often predict use, abuse, or relapse. With Internet and screen use, this is also a significant factor in that the Internet has become ever-present and easily accessible. Wi-Fi, high-speed Internet, smartphones, fast mobile service, tablets, and smart TVs have all brought the Internet to our fingertips. My prediction is that 5G (the next generation of mobile service networks) will only increase the addiction potential of the Internet and our devices, but especially the smartphone. The smartphone is essentially an *always-on* Internet portal that allows the Internet to be easily accessible, which I believe accounts for some of the Internet addiction cases over the last ten years.

Availability is like ease of access, but it speaks to the fact that Internet access is nearly *always* available, as smartphones, hardwired broadband home or office access, LAN, Wi-Fi, or access through your cell service makes the Internet nearly always available. More rural areas may be an exception to this unfettered access, but with satellite and fiber optics, this is also changing. In fact, many of us will soon rarely be without access to the Internet, which just increases the allure and temptation of checking that app or website, viewing that post, sending a text, or playing the latest version of your favorite game. It seems that there is no easy way to avoid easily accessing this digital drug.

REMEMBER

When an addictive behavior is easily accessible and available, it increases the likelihood of addictive use. It's also important to remember that a big part of why addictive use of Internet technology is problematic is because of the amount of time it steals from our lives without our being aware of it.

Talking about time distortion

Perhaps one of the most powerful drug-like effects of screen use is the time distortion that occurs online. One of the hallmarks of any substance or intoxicating behavior is that it can alter mood or consciousness, thus creating a time-distorting experience. In my 1999 study, I found that a large majority of users reported the time-distorting effect of the Internet. Perhaps you yourself have noticed losing track of time when you're online? The immersive and dopamine rewarding experience of the Internet clearly impacts our perception of the passage of time.

THE FIRST LARGE-SCALE STUDY OF INTERNET ADDICTION — AND WHAT HAPPENED NEXT

I worked my way through 12 years of college and professional training as an electronics technician, and I always had an affinity for science and electronics, so computers and the Internet fit right in. I believe it was this coalescence between my interest in electronics, addiction medicine, and human sexual behavior that offered a perfect combination for my newly developing specialty in Internet and technology addiction.

In the earlier days of the Internet, I collaborated with a colleague, Dr. Alvin Cooper, on some articles and publications on the interaction between the Internet and sexual behavior. Dr. Cooper was an early pioneer in research on the addictive nature of the Internet as well as Internet sexuality. Around this time, I read a paper written by the late Dr. Kimberly Young, who published the results of a small pilot study comparing Internet addiction to *pathological gambling* (the medical diagnostic term for gambling addiction). In her study, she identified several factors that seemed to play a role in the addictive nature of the Internet. I was intrigued, and shortly after, I embarked on a partnership with ABC News to develop and publish the results of the first large-scale study on Internet addiction.

Research in psychiatry and behavioral science is an arduous process and involves attracting subjects to study, so we decided to develop a broad and statistically sound survey that would be posted online for people to complete and that would give us an idea of what they were doing and experiencing online. There are, of course, limitations to a self-selected and self-reported study, but it was a first step. So, in 1997 the study finally went up on the ABC News servers, and in less than two weeks, we had over 18,000 responses. Some subjects were thrown out, but even with these rejections, this was an unheard-of response and gave us thousands of lines of data to look at, thus beginning an entirely new subspecialty of behavioral addiction medicine for me. I often tell doctors and clinicians whom I train that my specialty sort of found me, as opposed to me choosing it.

It took months to analyze the data, which led to a journal publication and resulted in my 1999 book entitled *Virtual Addiction*. At the time of my research, only one book had been published on the subject and there was almost no research; today, there are hundreds of books on the subject, and you can find nearly a thousand research studies and clinical papers on it. In short, we know a lot more about the Internet than we did in the mid-1990s.

The result of my original research (which has since been supported) is that yes, the Internet and the portals through which we consume it are clearly addictive. It is well accepted that Internet addiction and video game addiction are legitimate diagnoses, which, in 2019, led the World Health Organization (WHO) to declare *Gaming Disorder* an official medical diagnosis to be included in its upcoming ICD-11 (International Classification of Diseases, 11th revision), which is the leading manual for all medical diagnoses used in healthcare.

REMEMBER

What time distortion tells us is that the Internet digital drug is powerful and can alter our perceptions. The reason for this appears to be, in part, that the Internet and screens hyper-focus our attention while interactively distracting us — while transmitting stimulating and intoxicating content at the same time. When you couple all of this with variable rewards, you can see how a person can become stuck online and get lost in cyberspace.

TECHNICAL STUFF

Digital drug delivery is very similar to the way drugs are consumed and metabolized in our bodies. The shorter the time between introducing an intoxicating drug or behavior into our nervous system and subsequent intoxication, the more addictive that drug or behavior is. This is because our nervous system is very responsive to the brevity of time between stimulation and intoxication — in essence, this is classical conditioning. The shorter the time that passes between when you click and then see, find, or play something online that is desirable to you, the more addictive potential it has. This speed factor likely means there are potential negative implications to faster processor speeds and high-speed Internet access connections — in this sense, faster may not always be better.

Giving you the world online: The illusion of online productivity

The Internet seems to create an illusion of productivity. Perhaps this is in part because we use our screens so much for school and work (and during the COVID pandemic, even more so). The problem is that the Internet doesn't actually increase productivity when you factor in all the wasted and unproductive time we spend on our screens. This is not to mention all the glitches, lost passwords, Internet outages, and other technical difficulties that we experience. There are estimates that 80 percent of the time we spend online is *not* for productive purposes, and much of the time we spend online is related to surfing or scrolling around, social media, pornography, video gaming, sports, shopping, or other intoxicating behaviors that devour time and attention without our being aware of it.

The Good (or Bad) Stuff: Factors Involving Content

My research found that part of the appeal in accounting for the addictive nature of the Internet was an interaction between stimulating *content* and the *process* of Internet delivery of that content. It seems that the interaction between the attractiveness of any content can be amplified using the Internet modality. I call this *synergistic amplification* (a mouthful). It is as if online content were the drug, and the Internet delivery mechanism was the hypodermic delivering that drug to our nervous system. It is this amplification that gives the Internet some of its power and potency, as you find out in this section.

Finding out about content intoxication

What is content intoxication? This involves understanding how varying forms of content, when consumed via the Internet, have greater power to elevate mood and instigate changes in the reward center of the brain. Many of these forms of content, such as shopping, stock trading, gambling, video gaming, and porn, existed prior to the Internet and were quite pleasurable (and potentially addictive), but when consumed through the Internet medium, they become even more intoxicating and addictive.

For example, pornography has been around for quite some time and certainly existed before we had the Internet. It was always stimulating to look at it, when viewing it in books, magazines, films, photos, or on videotapes and DVDs. But when the Internet became more widely available in the early to mid-1990s, it was a new ball game. The ease of access, availability, anonymity, disinhibition, and convenience all coalesced to produce an immersive experience unlike previous forms of media or communication. The Internet allows for endless variety, privacy, and interactivity in a way never seen before. This was content on steroids, and thus began an entirely new type of addiction. (Chapter 9 has more information on pornography online.)

REMEMBER

The Internet amplifies the power of anything it broadcasts in a new and different way, and it happens without your knowing that your brain is slowly being conditioned — just as if you were playing a slot machine. Rewarding and stimulating behaviors are one thing, but when provided via an Internet platform, they become quite another.

Mixing stimulating content and digital devices

Why would a video game be more addictive when played on a computer or gaming platform connected to the Internet? In a word or two: *variability* and *interactivity.* The Internet allows people to play games with other people, and this, along with speed, variation, privacy, and ease of access, makes it entirely new. Digital devices, when connected to the Internet, make content much more personally accessible — to a point where the line between normal and excessive use can easily become blurred.

All forms of content consumed on the Internet are amplified and intensified. This is true whether we're talking about shopping, investing in stocks, gambling, video gaming, and most certainly pornography. Even searching for information becomes a new experience I call *infotainment*. Who would have thought that surfing YouTube, Reddit, Wikipedia, or a variety of apps or websites could be fun? But the interactivity, speed, perceived anonymity, and ease of access put simple information on *steroids*. The variability of finding what you want and not knowing when you will get it gamifies the process, and at the same time you're getting intermittent hits of dopamine when you find something you like, and even higher hits when you think that *maybe* you found or *will find* something you will like.

Understanding instant gratification

The Internet allows for a near-instant reflex, where a click or screen tap enables you to find just what you're looking for. There is very little lag between the impulse to look for something, play a game, or respond to a notification and the act of doing it. Keep in mind that the shorter the time between clicking and viewing the content, the more addictive it becomes. The shortened lag also produces a sense of instant or near-instant gratification and reinforces your inability to delay gratification. What this ends up looking like is that you never really have to wait for anything. Everything is instantly experienced, from a whim to satisfaction.

Sometimes waiting can build our tolerance to address aspects of life that are not instantly satisfied or that require sustained attention and effort. At times we might even learn to endure boredom for a few short minutes, without reaching for a screen, but the Internet seems to facilitate the opposite.

WARNING

Data seems to show that the ability of younger people to delay gratification and maintain sustained effort has waned over the last 25-plus years. Add to the equation that we're carrying an Internet portal everywhere we go, and we can see how our smartphone erases the last vestiges of our willpower by allowing us to instantly satisfy every impulse. Every social media update or notification we receive becomes a trigger to pick up our phone and look. Every question, curiosity, or text we have becomes another glance at our phone. The problem is that it never ends, and any boundary between screen life and real-time life evaporates. There is no off button, no downtime to enjoy without the pull of our phones and the lure of that instant dopamine hit.

TECHNICAL STUFF

Did you know that the anticipation of finding or seeing something desirable or stimulating (dopamine-releasing) is stronger than the actual pleasure itself? In other words, the *anticipation* of seeing something you *might like* will produce even *higher* levels of dopamine. Just like in gambling, it's the expectation or belief that you *might win* that is most intoxicating. So, if you post something on social media and see notifications come in, they will be more dopamine-elevating than looking at your phone and seeing that your post was liked or commented on. To your brain, being in the game is more powerful than winning the game — but winning also provides an additional secondary hit of dopamine.

Facebook uses this anticipation factor in the form of staggered posting of your received likes or comments, and then delivering them to you *randomly* to keep you looking at your page over and over.

Defining infotainment

Early on in my work with Internet addiction, I coined the term *infotainment* to describe what we all do to some extent with the Internet, and especially our smartphones. Information has taken on an entertaining quality, and it can keep our eyes onscreen for content providers to data-mine and to sell us things. I would add that social media has this quality, driving our excessive use. This is new stimulation to our brain, and our limbic reward center is scanning for novel and stimulating pieces of information.

Digital drug delivery is very similar to how drugs are *metabolized* in our bodies.

This Must Be the Place: The Internet as the Car, Map, and Destination

The Internet is a unique form of technology; it's not just a communication or information tool, but it's also a *place to go* in and of itself, as you discover in this section. Early in its history, we assumed that the Internet was like many earlier communication modalities such as the telephone, radio, movies, and television, and that the Internet would simply be another tool to connect and entertain. Although e-commerce was seen early on as a possible use for the Internet, no one could have imagined how quickly it would morph into not just a means to connect, but a *destination* to connect to as well.

What do I mean by a destination? The Internet is a place to virtually hang out, not just on chat sites, but to shop, trade stocks, game, gamble, or hang out on a social media site like TikTok, Reddit, Twitter, Snapchat, Instagram, or Facebook. The Internet is not simply a means to an end, but it has become an end to itself. It was a new place to go, and it has only increased in this capacity to entertain and captivate. In fact, there is such a small boundary between real-time and virtual these days that many of us don't differentiate between the time we are online and the time we are off-line (even though essentially, we are never off-line anymore with our smartphones always in our pockets).

Unlike when the Internet made its debut and we had to dial up to get online (kind of like getting in the car to go somewhere), the Internet connection is now *always* live with high-speed DSL, cable, satellite, fiber-optic, and fast smartphone connections. Even a slower DSL connection is always on, and a seamless highway is always open on your desktop, laptop, smartphone, or TV. This connected factor is in part what makes the Internet so powerful — and there is something intoxicating about having the world's information and people at your fingertips.

Getting the word in and out: Broadcast intoxication

One of things we discovered early on with the Internet was *broadcast intoxication*. This is essentially a recognition that it is stimulating to broadcast to others (as well as receive broadcasts) through various platforms, but especially on social media. In Chapter 5, I discuss in detail how social media creates an intoxicating experience when you broadcast your status, updates, posts, photos, videos, or virtually anything. Likely, this intoxication is in part caused by the anticipation of a *like, comment, follow,* or *DM* (direct message) to your post. The response by other people creates a social validation loop whereby there is a dopamine elevation from the social validation. Part of the reason for our intoxication from this validation is that we are all hardwired for social connection and social approval. The Internet, and especially social media, may capitalize on a very basic human quality: the desire to be liked and appreciated.

Weaving a web: A story without an end

All forms of communication and media have boundaries. A book, TV show, newspaper, magazine, movie, and even a text or phone conversation all have boundaries. They all have a beginning, middle, and end, and there are markers that tell you where you are in the entertainment, information, and communication process.

The Internet, however, is an entirely different matter. Whenever you go online to do anything, there are no markers for where you are and how long you might be there. In fact, there is a purposeful attempt to eliminate such markers, just as a casino removes clocks or windows to obscure time passage. There are a myriad of cross links, back links, hypertexts, live photo links, click-bait, and feeds that take you down endless rabbit holes, which have you later emerging from your journey without a clue as to how this occurred. *No boundaries equal no markers for time passage.*

The Internet is a completely dynamic, active, and interactive system. It isn't linear, but rather a networked and almost circular set of interactive data. It in some sense operates more like our brain than a book. But without markers for time or space, the Internet can take you on a journey far beyond where you intended to go. Obviously, online content providers and Internet companies love how this lack of boundaries creates captive audiences of you and me.

REMEMBER

The Internet and digital screen technologies are amoral. They have no agenda in and of themselves, and service providers, content developers, and app and software creators all have the same goal: your attention and your eyes onscreen. This is a battle for how you spend your time, but your time is a non-renewable resource, and technology must always be balanced in terms of how it ultimately serves us, not simply for a promised better life.

Apprehending the myth of multitasking

There is perhaps no better example of the purported benefits of screen technologies than the power of multitasking. Multitasking is a very misunderstood concept, and there has been much research on the neuroscience of attention; the overwhelming conclusion is simply that there is no such thing as multitasking.

WARNING

What is often seen as multitasking is rapid *attention-shifting* and moving your focus in an alternating fashion. What this means is that you're quickly shifting your attention from one screen or activity to another, and although it feels seamless and simultaneous, it isn't. There is no way for the human brain to attend to and process two stimuli at the same time. So, the next time your teenager has a laptop, tablet, TV, textbook, and smartphone open in front of them (while they are listening to music) and then tells you they are attending to all of them at once, they are kidding themselves (and you!). They may believe they are attending to all these activities simultaneously, but what the research shows is that the amount of comprehension of each stimulus is basically reduced by a factor of how many other sources of input you have going on at once. So, if you are doing homework while doing other activities, it will simply take you longer to get it done and/or there will be less comprehension of what you worked on. Internet and screen technology does not actually increase efficiency; it increases the functional organization of how we manage information, but even this benefit must be weighed against the amount of *distraction* it creates along the way.

Telling a social story: The net effect on people

The big question is this: How does Internet and screen technology affect human relationships and our overall health and well-being? From its early adoption, the

Internet has been touted for its ability to connect people; initially social media was hailed as an important way to stay connected with our friends, family, work, and school. There is little argument that the Internet is a useful tool, and the COVID pandemic has offered further evidence of how amazing and useful this technology is. It has allowed us to continue our work and school from home and to shop and stay connected to our friends and loved ones. I am in no way debating the utility of this technology; what I am addressing is the quality of some of those activities and social connections made online.

There is little disagreement as to the mixed quality of online social connection and questionable satisfaction (and efficacy) of attending school online. Why is that? Why is it that social behavior online seems decidedly two-dimensional? Why has online schooling during COVID been seen as a failure? Why is online intimacy and social connection generally rated as inferior to real-time interaction? Even Zoom, which may be one of the heroes emerging from COVID, can leave people feeling hungry for more real-time, physical contact and connection. From the COVID experience, we see that our children, who have been schooled virtually, report increasing depression, social isolation, and dissatisfaction from the online academic experience — and some are experiencing other increased mental health issues as well.

These experiences aren't a coincidence. They are reflections of some of the weaknesses of Internet interactions for social and academic communication. The Internet is best at connecting people with information and services, but it leaves people yearning for more depth and breadth when all is said and done. Perhaps it is best said that our expectations for a two-dimensional experience may be higher than what an Internet screen can deliver.

The Human Factor: The Internet as a Digital Drug

People sometimes want to know how the Internet and digital screens can be addictive. After all, how can you get addicted to a screen behavior? Part of the reason for this question is that it evolves from the idea that addiction is about ingesting a substance and that it is the *substance* itself that creates the addiction through the body's physical dependence on it.

REMEMBER

The problem with this analysis is that it's wrong. This isn't how addiction really works. Yes, the body can become physiologically dependent on a substance (drugs or alcohol), but if that is what solely created an addiction, then once people detox from the substance, the addiction should be gone. However, this is not what typically happens. Addiction is the combination of many variables that involve learning, memory, emotions, social factors, physiology, behavior, and neurobiology. Often there are co-occurring psychiatric problems that contribute to an addiction, or the addiction may be a way to deal with emotional pain or negative circumstances. Let's face it: Addictive behaviors can be an escape.

With addiction, there is always a *disruption* of the reward system in the brain, but the addictive substance or behavior (in this case, Internet use) is not really the primary issue. It is certainly a factor, but the addiction process is as much about learning to deal with triggers and to manage one's emotions as anything else, and what we see with excessive screen use is like what we see in many addictions: Addictive behaviors start out as a solution that then becomes another problem.

WARNING

Being online is a pleasurable and stimulating activity that impacts the brain's reward system and elevates dopamine; it has a similar potential to create an addictive experience as drugs and alcohol, although physiological dependence is less of an issue.

Grasping the power of "maybe"

The Internet is the world's biggest slot machine. The slot machine operates on the power of *maybe,* and this explains the neurobiology of gambling, and most addictions, in some ways. So how does *maybe* work in our brains? When you pull the handle of a slot machine, your brain knows that it is going to *win* something. What it does not know is *when* and *how much,* but it *does* know that eventually there will be a reward of some kind. It learns to expect this reward. This system of reinforcement (or dopamine reward) is called *variable ratio reinforcement.*

So why wouldn't you just give someone a win each time they pull the handle or after they pull it a certain number of times? The answer is simply *boredom* (also called *extinction*). B.F. Skinner did groundbreaking research on operant conditioning and found out that unpredictable (variable) rewards create longer-lasting response habits (which is another fancy way of saying addiction). If a reward is given variably and unpredictably, then your brain engages in this activity like nobody's business.

So how does this work with Internet use? Whenever you go online (or on your smartphone), you never really know *what* you're going to find or *when* you're going to find it. This is true for email, information surfing, scrolling, gaming, social media, shopping, and porn; virtually anything you search for online has a variability to it — a *maybe* factor. In essence, the Internet operates on the same variable reinforcement schedule that a slot machine does, and people are neurobiologically conditioned by their devices without really knowing it.

Not only does the brain love *maybe,* but it also loves *newness.* Novelty resets our attention and interest. And there is perhaps no better source of information that provides endless variability and novelty than the Internet. Every time you go online (and in a sense, we are always online with our devices), it can feel like a new experience each time you click or tap through your latest impulse.

Seeing how dynamic interaction keeps people coming back for more

Dynamic is a way of saying changeable and interactive — and being able to interact with our devices puts some psychological skin in the game for us. It's dynamic in that we are driving the interactive process by our clicks, and the cycle is complete when we are *responded* to by our screens. We self-guide our web adventure, and regardless of whether we are on our smartphone, laptop, tablet, or other device, our Internet journey feels like a dialogue. The AI (artificial intelligence) interfaces by Amazon and Google provide a voice to this interactive exchange, giving the cyber experience a very compelling feel. We now experience the Internet not simply as a source of information, but as having lifelike qualities, and the line between simplistic Internet algorithms and AI is fast eroding. I frequently laugh to myself as I enter discourse with Alexa on my Amazon Echo. I know full well that she is just a machine-learning interface, but she feels real and is thus imbued with the dynamism of humanity.

REMEMBER

Most of us would be hard-pressed to admit that we love being online, but the fact is that many of us do. Some of us may find ourselves becoming addicted to a point that our lives are impacted, while others of us may overuse at times. As I mention in earlier chapters, much of this is *not* conscious. We are virtually swept away in a tidal wave of intensity, stimulation, interactivity, and endless intermittent reward. The problem with tidal waves, however, is that they can hurt us, and our excessive Internet use can subtly rob us of many aspects of quality living. In more extreme circumstances, once addicted, we see our lives becoming impacted in negative ways that we did not fully appreciate. It is this lack of awareness that blinds us to the slow erosion of our most cherished values, and excessive Internet use can mindlessly peel away our time and attention to a point that we forget what real-time living feels like.

2

Breaking Down Addictive Technologies

Define the addictive nature of social media and how social validation looping keeps you endlessly viewing, posting, and commenting on your social media apps and websites.

Check out the growing phenomenon of streamed online entertainment and the addictive nature of consuming content in this growing Internet-based format.

Recognize the addictive nature of video gaming and how games are designed to capture more than your imagination. Nearly three out of four households in the United States have someone who video games — that is, according to 2020 statistics, about 214 million people — and for most, it's a fun form of entertainment. However, a small percentage of people find themselves lost in their game world without finding a way out.

Understand that the Internet fuels the power of purchasing and gambling online. Online shopping, gambling, and stock trading are very different than when they're done in other formats, and present unique challenges in not crossing the line online.

Examine the addictive nature of Internet pornography and the long relationship between pornography and technology, which, when woven together, create a powerfully addictive combination.

IN THIS CHAPTER

» Breaking down the phrase *social network*

» Demonstrating the power of social validation looping

» Checking out social media's effects on communication and self-esteem

» Looking at the counter-social aspects of social media

» Finding relief from social media

Chapter **5**

Examining the Addictive Nature of Social Media

What exactly is social media? We scarcely go a day without hearing the term, and it seems everyone is using Facebook, Instagram, Snapchat, Twitter, TikTok, Reddit, YouTube, LinkedIn, and endless other platforms. Social media has invaded almost *every* aspect of our lives, including shopping, politics, data collection, entertainment, influencing and marketing, news, and — in perhaps some ways — social connection.

Social media was initially created to connect people online. It has since morphed to including topics, interests, and being part of a local or wide "friend" network; photos, videos, and other personal information can be shared, rated, liked, and commented on. Facebook (currently with 2.85 billion users worldwide) was one of the earliest social media platforms, but others such as Myspace share an even earlier pedigree. More recently, social media has morphed into a colossal media machine and e-commerce platform, giving birth to social economics and social commerce, where advertising, sales, and marketing are all integrated.

For many users, social media, in all its iterations, has become a way to communicate regularly with friends and family by posting and scrolling through posts to view news, photos, and information about their lives and often to pass time. Social media represents a huge amount of time on our screens, and advertisers and marketing companies don't waste a second of that social capital. Although many social media websites and apps have the proposed goal of connecting people in some way, the quality of these connections is ultimately questionable. Is social media simply a convenient shadow of real-time social connection? I think the truth is probably mixed, as there is no doubt that many people experience connection and support from social media, but whether this can serve as a comprehensive substitute for real-time social interaction seems unlikely.

Typically, social media websites and apps are free to users (at least, on a direct monetary level); the companies typically make money through advertising and sales of data and metrics about you. If you doubt the financial profitability of social media sites, just check the stock market prices on some of your favorites, and you'll see just how profitable they are.

The initial impetus behind social media might have been to connect users, but it soon became apparent that it was necessary (and desirable) to financially support these services by capitalizing on the one key asset that social media companies have — you. Your eyes-on-screen were the currency that ultimately became the funding to support social media sites. The illusion that social media is free could not be further from the truth. If something online ever appears to be free, then your attention and demographic data pay for the service with the most precious commodity of all — your time and attention.

In this chapter, I dig into social media's addictive nature, its effects on communication and self-esteem, its counter-social attributes, and much more.

A Social Network: A Rose by Any Other Name

I want to take a moment to dissect the phrase *social network*. In essence, what we have is a network of interconnected online users who engage interactively in some fashion. This may seem like a loose definition of social connection, but if you visit any social media site, you can see that the level of social *interaction* seems somewhat superficial, and I contend it is a poor substitute for real-time social interaction.

This is not to say that people do not connect via social media platforms (can you imagine not having Instagram, Facebook, or other platforms during the COVID pandemic?). Today's youth, starting in grammar school to well beyond their college years, utilize social media systems to communicate and express themselves to their peers. The fastest-growing segment of social media users, however, are adults (even older adults), and as tweens, teens, and college-age kids have abandoned more established social media sites such as Facebook, they have migrated to other newer sites, such as Instagram, Snapchat, TikTok, and others.

REMEMBER

It's important to note that social media was originally intended as a *social connection platform*, but at this point it's difficult to separate the social and economic aspects of these sites. As social media has become increasingly monetized through ads, sales, and marketing, it seems there is less emphasis on social community and more of being on a marketing platform. I would even question whether the term *social media* is misleading, in that the only reason why these apps or websites seem to thrive is because your profile and viewing activity are potential revenue sources.

Recognizing What Makes You Come Back to Social Media for More

Imagine having a conversation with a friend; she is leaning forward, moving her head in response to your words, showing facial gestures that are indicative of someone who is listening and engaged in what you're saying. You behave similarly in return, and this interaction becomes the basis of the natural ebb and flow of communication. This is an example of *social connection,* and we are hardwired biologically to positively respond to such interactions. Social connections, such as what I describe here, are highly rewarding, pleasurable, and health enhancing — especially when sharing is mutual. And because these experiences are pleasurable, these interactions are reflected by an increase in dopamine in your brain.

Social media operates much like this, but instead of real-time conversation, we use posts, instant messages (IMs) or direct messages (DMs), comments, photos, and videos that are responded to with *likes, comments,* and *follows.* In the next sections, I explain two concepts in more detail: *social validation looping* and *variable reinforcement.*

TECHNICAL STUFF

Some evidence suggests that the use of alcohol and drugs has decreased in recent years for adolescents, and Dr. Nora Volkow from NIDA (National Institute on Drug Abuse) noted that this could be occurring because screen use may be providing the dopamine hits that were previously obtained from substance-based intoxication. In some sense, social media and screen use may be providing some of the fun and excitement previously obtained pharmacologically. Our screens, and in particular our smartphones, offer easy access and availability, and therefore can serve to "medicate" moods, emotions, and frustrations. It's like carrying a portable dopamine pump in your pocket, and we know that the more readily accessible a substance or behavior is, the more likely we'll use and, at times, overuse it.

Looking at social validation looping

REMEMBER

Every time we get a *positive social reinforcer* such as a *like, comment,* or *follow,* we tend to receive a small hit of dopamine; the fact that these positive social media responses are unpredictable and variable enhances the potential to produce a pattern of repetitive behavior. We are *rewarded* for our posts (through our likes, comments, follows, shares, and retweets), and so we post and check *again* to obtain *more* positive responses. Each positive response we receive elevates our dopamine level in a small way, and the unpredictability of the responses we get keeps us coming back again and again. This dopamine elevation is essentially based on the naturally rewarding (and hardwired) aspects of feeling socially validated and desired. This process is referred to as *social validation looping,* where there is a cycle of expression (posting) followed by responses (likes, follows, and comments), again followed by more posting, which starts the cycle over again — hence, social validation looping.

It becomes even more insidious because even the *anticipation* of a positive response from others can elevate our dopamine level; posting alone can *cue* the reward centers in our brain to anticipate a response, and this cycle can continue indefinitely if we get an occasional reward in the form of a *like, comment,* or *follow.*

We use the term *social validation* because social media providers are utilizing our inherent desire for social interaction and social connection — to be validated socially — to shape our behavior on social media apps and websites. The word *looping* refers to the fact that this behavior creates a repetitive cycle. Everyone feels good when what they post is acknowledged, and the likes and comments let us know we are being appreciated and valued. All these forms of social validation from our social media posts *variably reinforce* us in powerful ways. The likes, comments, and follows that we end up with produce that same slot machine–like addictive phenomenon that we see with many aspects of Internet behavior. Social validation loops are what motivate the endless wheel of posting and checking for *likes* and *reactions* and then posting again and again.

TECHNICAL STUFF

Part of what makes our brains so efficient and adaptive, from an evolutionary biology standpoint, is our ability to organize and categorize various forms of input that we then recognize as patterns; this allows us to predict potential outcomes that help increase safety and efficiency (thus increasing our survival potential).

REMEMBER

If our brains *neurobiologically expect* certain behaviors will be rewarded, it is an efficient shortcut to anticipate the outcome and consequence. All addictive behavior has its origins in the hijacking of survival-based reward pathways, so addiction has a functional place in our biology, even if it can ultimately lead to negative consequences.

Understanding the big deal about variable reinforcement

Any habit or behavior that is reinforced (rewarded) intermittently and unpredictably will last a long time because the brain likes *maybe*. If you have any doubts, look at the lines of people buying lottery tickets, watching TikTok, sitting in front of slot machines, betting on athletic events, investing in the stock market, or even surfing the Internet. Watch how people use eBay or other auction sites — it's the chase, the possibility, the *maybe*, that makes all of this appealing. *Variable reinforcement* is just what it sounds like: When our behaviors are intermittently reinforced in some way, they are much more likely to continue for long periods of time. We call this *extinction resistance*, and it's what helps create a *habit* or an *addiction*, depending on circumstances and life impact.

It is my understanding that in the beginning, Facebook shares and posts were routinely made available for the public or *friends* to view. It worked well, but it was not until one Facebook employee created the *Like* button that things really took off. People began to post like crazy (no pun intended), and more posting meant more time onscreen. The *Like* button was the critical ingredient that was needed to produce social validation looping (see the previous section). Getting those *likes* (especially variably) really made users pay attention to their profile and the site and led them to keep checking it over and over. Doling out *likes* and comments in a delayed and intermittent or random format takes further advantage of the power of variable reinforcement. Soon after Facebook installed the *Like* button, users began posting and checking to see what people *may have said* about their latest post or comment, and this was a social media game changer.

REMEMBER

Social media creates the impression that living life is not rewarding in and of itself, but rather it is the *recording* and *sharing* of your life that makes it meaningful. (See the nearby sidebar "The risk of selfies" for details on a popular method for sharing your life.) Posting your experiences via photos, words, and video allows others to respond to your life and, in a sense, creates a form of *reflected self-esteem.* You feel good about what you did, not because of your experience, but because of *how others rated* your experience. You are no longer simply living and experiencing your life, but rather, you are experiencing it as a form of validation from others. This cycle only leads to disappointment, as you always end up comparing yourself to others in the process.

WARNING

You can never achieve self-satisfaction when using *social, material,* or *physical comparison* as your basis for self-esteem. Social comparison, which social media apps base their popularity on (especially for younger users), allows for superficial comparison of physical and other attributes of success as a valid reality and means of deriving identity.

THE RISK OF SELFIES

Taking photos of oneself — *selfies* — and posting them on social media is a very popular behavior among teens and young adults, but it may be practiced by younger children and adults as well. In essence, it is the same behavior no matter who is doing it.

Taking a photo or video (preferably an interesting one) and posting it is all about the *likes, comments,* and potential *compliments* you might receive once you post it. In extreme cases, this can lead to almost obsessional levels of photo-taking activity. There have been numerous instances where injuries and deaths have occurred in the process of trying to take the best and most exciting selfie possible, as sometimes these photos involved risky locations and situations. The desire to achieve likes, comments, and positive feedback drives people to take risks to get the best shot possible. In some cases, great care will also be taken to touch up photos to make them even more perfect.

The sad irony here is that all this effort to achieve external recognition offers limited real benefit. The net effect (there's that pun again) is to keep trying to capture more and more novel and exciting photos in order to continue to achieve the desired effect and to receive external validation (another example of social validation looping).

Taking selfies can be dangerous to your health and can even kill you. In 2019 there were a reported 259 selfie-related deaths (and these were the reported ones). Many of these photo ops were taken in obviously dangerous situations that went wrong. Taking a photo hanging off the edge of the Grand Canyon is never a good idea.

The problem is that most people *do not* necessarily post accurate, honest, or complete portrayals of themselves (many users alter their personal images) and typically post only the best photos, videos, or circumstances that further propagate the idea that *life is the way it looks on social media.* However, nothing could be further from the truth. Social media portrays a two-dimensional view of who we are, and often leaves out our most human and humane qualities. Children, tweens, teens, and young adults are the most susceptible to these distortions and to unrealistic social comparison, but adults are susceptible, too.

Seeking Communication and Self-Esteem — But at a Price

Can you use social media to connect, build, or maintain social intimacy? Well, this is the question of the day, and the answer is mixed. Some studies have looked at how social media use translates to deeper and more conventional aspects of socialization and relationships. The best we can glean so far is that social media is a very different form of communication and may in fact represent a completely new type of social interaction.

Having used it myself, I can attest to its initial attraction, fun, and usefulness, but it always feels somewhat utilitarian and flat to me. Social media leaves me feeling ultimately dissatisfied, as it lacks the depth and nuance of real-time interaction. That is not to say there is no place for it, but I question how much time and energy we need to spend on it, relative to other forms of social connection. At this point in time I rarely use social media, with possible exceptions for work or if I'm asked to view something from a friend or family member. Some of the reason for this is that I can feel its addictive pull and see how easy it can be to get caught up in the cycle of posting, waiting for a reaction, and then posting again — or simply *scrolling* endlessly.

The Internet is a new medium (relatively speaking), but frankly, when the telephone was invented, people initially thought it would have no usefulness. Now, with texting, chatting, IMs, DMs, Snapchat, Twitter, and other forms of social media, some might say we really have no need for the telephone, but there is probably room for all modes of communication. I am fairly convinced, however (and the data seems to support this), that social media is *not* a replacement for real-time social connection.

Because most social media interactions utilize written messages as well as photos and video clips, there seems to be a difference in how messages are transmitted, received, and ultimately encoded in our brains. Typed messaging allows more distance (but less nuance) than does verbal communication, and the one-way and delayed effect of a photo or video message is quite different from an immediate, interactive video chat using an app such as Zoom, Skype, or FaceTime.

Research seems to support the idea that social media communications may lack sufficient social cues and nuance, and that if social media is the dominant modality by which you communicate, it can produce a perspective that lacks social empathy and depth. This may, in part, be due to the long-distance way in which information is shared — even if that information is of a personal nature. I characterize social media communication as a less *nutritive* form of connection, but frankly, we do not have any long-term studies that look at how social media use impacts interpersonal and social development in the long term. The preliminary evidence suggests that there are some deficiencies relative to this type of communication. Social media technology is still relatively new, and we have only about two generations who have been raised on it and who are just now entering mature adulthood. Newer iterations of social media will undoubtedly be different, but one thing is for sure: These companies will need *your eyes onscreen* to monetize their platforms.

WARNING

The superficial aspects of social media, when it is used to communicate to a wider and larger audience outside of immediate friends and family, leads to an interesting phenomenon. The idea that you can attract a large cadre of followers and that you can influence people (or be influenced by them) may have its own intoxicating effects, with or without potential economic rewards. The satisfaction and excitement associated with having *likes, comments,* and many *followers* on any platform of social media is notable, but it is questionable that it offers any ultimate value beyond ego validation. There are numerous examples of YouTube, Instagram, and TikTok stars who are famous within these mediums. This may be akin to being a popular person at school or work, but my concern is that this lack of depth can lead to a risk of supporting superficial social exchange. It is not either/or, but it seems important that a balance be achieved.

Self-esteem that is connected to the perception that one's posts and communications via social media are well received, or received at all, is potentially problematic. There is an inherent risk that you can post something and have *no one* see it, or worse, that people will read it but *not* like or comment on it. The token of social exchange within the social media world is *some reaction* that is then communicated back to the user who initially posted it. This interaction can become a two-way social exchange, albeit with some delay, but it is not a guarantee.

WARNING

When self-esteem is tied to how your audience or online friends react, this leaves you vulnerable to only using *reflected* assessments in deriving your self-definition. Social interaction is a significant aspect of creating your sense of self, but it is questionable to what extent social media interactions can provide deeper (and psychologically necessary) levels of human connection; the term *social-lite* may be an apt description of such interactions.

REMEMBER

The best that might be said about social media is that it has produced a new form of communication, and just like other new forms of communication in the past (such as the telephone, radio, movies, and television), there are always adaptations and growing pains when fitting them into balanced living. The difference with the Internet and social media is that they can clearly produce addictive behavior, whereas other modalities don't seem to possess that level of power. Addictive behavior operates on a continuum, and many people use social media, as well other aspects of the Internet, with little or no ill effect.

WARNING

A fact that seems to be well established is that social empathy does not appear to be enhanced by social media. Social empathy scores are lower among heavy users of social media. The reasons for this are unclear, but it seems reasonable to assume that the cues for enhancing social empathy are somehow blunted when receiving information through social media modalities. This has significant potential implications when it comes to cyberbullying or trolling, and other negative uses of social media that shame or negate individuals via a social platform.

Seeing Why Social Media Can Be Counter-Social

In this section, I am not using the term *counter-social* in a traditional psychiatric sense. Rather, I'm saying that social media communication may run contrary to facilitating intimate connections. Clearly, social media *is* a form of communication, but my argument is that it doesn't seem to live up to its name, as it is only *sort of social.*

There have been many anecdotal reports of individuals who have discontinued their use of social media and reported feeling a sense of relief. They described feeling enlivened by getting out of the validation loop of posting, reading, and posting again. If you or a loved one uses social media regularly, you may be on a merry-go-round of posting and checking (all the while getting caught in the whack-a-mole of ads, cute cat videos, and a myriad of other distractions) and are trapped by the addictive nature of social media. Having removed myself from most social media on the consumer level, I would affirm that it is indeed liberating to not have to worry about an endless cycle of social media use.

Broadcast intoxication on social media

Broadcast intoxication, discussed in Chapter 4, is defined as the dopaminergic rewarding aspects of self-expression via the Internet and social media. Social media use carries with it many potential negatives, but the largest by far is how quickly we can become addicted to intoxicating effects of constantly posting and sharing our lives online. It promotes the illusion of social connection and intimacy, but in a buffered version. What we share via social media platforms probably reflects more of *what we want* the world to see about our lives and often consists only of socially desirable experiences. In other words, what we post is more a reflection of what we *think* may be interesting to others (and hence *likeable*), and might accumulate positive reviews, comments, and *likes.* The idea that social media is an expression of who we really are may be inaccurate because who we are on social media is at least partially shaped by the reactions we receive to what we post.

WARNING

The real challenge of social media is in learning how to be ourselves and to experience our daily lives without the impact of how our experiences might appear once we recount them to others via social media. Social relationships ideally promote not filtering what we do through how others will react and rate our experiences; this further disconnects us from ourselves and prevents us from experiencing our lives in the present moment, without the drive to photograph, video, and post our lives. *It is difficult to be fully present in our own experiences when we are evaluating how others will react to them.* The separation between living our lives and broadcasting ourselves has narrowed to a point where we have lost some of the ability to experience life *without* sharing it for potential review. This cycle is addictive and separates us both from ourselves and, I would argue, from real-time connection.

This is not to say that social media is all bad, but rather that social media is a potent time-sink and can easily suck you into a rabbit hole, eating your time, energy, and ability to live a balanced life. The allure of staying within this cycle is powerful, but the freedom that you can experience by breaking this pattern can provide that much more time and energy to devote toward internally gratifying pursuits and rewarding real-time relationships. It is my belief that this technology can *shallow our relationships* and separate us from ourselves, as the paramount focus is instead on the reactions of others.

Cyberbullying

When it comes to screen use, and especially social media, the adage should be *convenience* is the mother of invention, not *necessity*. The power of social media, and for that matter all things Internet, is the power to connect people to information as well as to other people. The purported purpose of social media is connection, but the reality is that large amounts of time are wasted and destructively expended. Cyberbullying is one of those dysfunctional and counter-social (perhaps antisocial) uses of the Internet.

This power to connect people, or to express thoughts and feelings about others, is what cyberbullying is about — albeit in a negative manner. There is a perception of anonymity when communicating online, and our original research, going back to the late 1990s, showed that people often experienced the Internet as anonymous. *When communicating online, there is a perception of distance both from those you are communicating with, and from those you are communicating about.* This perceived separation insulates us from the impact of what we say, and how what we say might affect others. In addition, there is no *social threshold* to cross, and no one's eyes to avert. The electronic distance allows us to say things in a more *socially disinhibited* manner.

Emboldened by social distance and the efficient bully pulpit that social media provides, cyberbullying has become all too common. Social media is a safe place to speak the unspeakable and say the unsayable, and these platforms often involve children and adolescents who are vulnerable to the powerful impact of peer-based opinions. The 2019 Youth Risk Behavior Surveillance System (a survey conducted by the Centers for Disease Control and Prevention) indicates that an estimated 15.7 percent of high school students were electronically bullied in the 12 months prior to the survey. Fifty-nine percent of U.S. teens have been cyberbullied or harassed online, according to the Pew Research Center (2018).

The potential negative impact of this power can be devastating. Bullying is always impactful, but cyberbullying may be even more so; there is an indelible footprint of every word uttered online and a record of everything said. Even if not on the front of our formal social media feeds, it is still always searchable, and some bullies will elect to save these posts to allow them to resurface another day. For those who are bullied, there is little defense. How can you defend yourself from a disembodied cyberbully who can say anything about you with an audience of your peers? The psychological impact is very real and damaging, often requiring psychiatric intervention in the form of counseling or therapy to help manage it. In some tragic circumstances, suicides have occurred because of cyberbullying.

It is all too easy, with the press of a button, to say things that cannot be retracted and that leave an indelible mark in cyberspace and on the psyches of those who fall victim to this prevalent behavior. Part of the Internet's biggest strength, its functional simplicity, allows the damaging, antisocial aspects of cyberbullying to flourish.

The footprint of anything written on social media is essentially indelible; its effect is immediate, and in some cases the damage can be permanent.

Cyberstalking and trolling

Cyberstalking and trolling the Internet and social media platforms is a common activity. There are estimates that upwards of 50 percent of the demographic and identifying information present online is a lie, and our original research found these numbers to be consistent with the 17,000 people we surveyed.

What is cyberstalking and trolling? It is basically what it sounds like. It's a virtual version of following, monitoring, viewing photos and videos of, and attempting to communicate with people online or on social media. Typically, this monitoring or communication is unsolicited and unwanted and can be very frightening and upsetting to the victim. Often, undesired communications can severely disrupt the victim's life, again resulting in needed counseling or therapy.

It is exceedingly easy to cyberstalk or troll someone on social media or the Internet because some information always leaves a public trail that can be followed or trolled. My best recommendation is *not* to use public settings for any of your social media, but unfortunately many teens and young adults (and fully grown adults) leave their profile open (and public) to make it easier for people to find and follow them.

Finding Relief: Life beyond Social Media

Perhaps some of the most convincing evidence to support using less social media comes from those individuals who stopped or greatly reduced their use, and the sense of relief and increased time and energy they experienced. Most of them were reluctant to let the habit go, and all of them felt uncomfortable and ill at ease when breaking the *social validation loop* (see the earlier section "Looking at social validation looping"). However, after several weeks without social media, most, if not all, of those interviewed felt a sense of relief and renewed energy to pursue life. No one expected this, but in story after story, even though getting off social media was often promoted as a temporary experiment, most decided not to return to regular use.

Ardent social media users need content (often the content is themselves) to attract the *likes* and *comments* (and possible financial rewards) they desire, which then provides the pleasurable dopamine hits; unless they are a content creator, their broadcasted lives become the content, and all consumers of social media pay for access to social media with their time and attention. Social media is now a dominant force in influence marketing, social economics, and the social e-commerce business, and our eyes-on-screen are the currency that keeps this engine running.

The illusion is that social media is somehow about the user, and perhaps to some extent that is true, but the reality is that companies that develop and market these sites and apps *are making money from you* either directly or indirectly. This is not to say that one cannot experience a sense of community and connection with those they interact with on social media. After all, social media is a vehicle that uses the power of the Internet, which offers the ability to connect mutually interested consumers of information with a supply of that information; social media can excel at this in that social activities, business, interest groups, dating, and social and cultural activities all have a home on social media platforms.

REMEMBER

Much good can be done in connecting like-minded people who share an interest or circumstances and need a platform to facilitate the self-disclosure process. Care should be taken, however, in that social media sites can be quite compelling, and all things Internet can eat up your time. Choices must be made in how you spend the *one commodity* in your life that is essentially priceless (and irreplaceable) — your time. Social media has the capacity to entertain, sell products, and in some ways, connect people. Head to Part 4 for pointers on living a balanced life, with the Internet and technology in its proper place.

Chapter **6**

The Endless Stream: Binge Watching TV and Online Entertainment

Television has really changed in the last decade or so. You've probably heard in the past about the impending transformation of television and how it's going to become interactive, but it wasn't until Netflix's streaming revolution in 2007 that a confluence of factors occurred, finally cementing the marriage of the Internet and TV.

The advent of user-friendly smart TVs and aftermarket devices like Apple TV and Amazon's Fire Stick to access streaming apps and alternative program sources has been a game changer — and more accessible broadband Internet provided the final ingredient for the streaming formula. For better or for worse, these changes have allowed television to finally be free of the constraints of scheduled programming, where you built your viewing around what the networks or cable companies provided. Although you could previously record your programs with a VCR or DVR, the ability to watch entire seasons of TV programs all at once, at any time, in any order, is a new phenomenon. However, these innovations have produced darker screen issues in the form of binge watching. Like all things consumed via the Internet medium, television is amped up to potentially addictive levels when integrated with the Internet.

A *binge* is the repeated and unconscious use of something that may have a detrimental effect on you. You often hear about binge eating, where a person eats compulsively, often to a point of hurting themself. The same might be applied to binge TV watching. The ability to turn on your TV and stream ten episodes of an already-released season can be very appealing at best but negatively impacting at worst — the main negative impact being loss of sleep and not getting other things done.

You can't binge without something powerful shutting down your conscious judgment and awareness. Only potent and addictive substances or activities can accomplish this. Pleasurable activities can trick your brain into experiencing what you're doing as good for you (enhancing life and survival potential) when, in reality, these are potentially addictive behaviors that are stealing your valuable time and energy. In this chapter, you discover the pitfalls of watching too much TV and what you can do about it.

Missing Your Life While Being Entertained: The Ease of the Binge

There are perhaps two clear saving graces that the Internet and screens have given us during the COVID pandemic. The first is Zoom, which allowed people to socialize, work, and attend school virtually, and the second is streaming entertainment services such as Netflix, Amazon Prime, Hulu, and others. These services have provided people with entertainment solace at a time when they *really needed the digital distraction.* I view these as wins for Internet tech, and they are good examples of the many benefits that online communication and entertainment can provide.

However, prior to this unexpected need for online content, people had other issues with their tendency to consume too much of a good thing. The *streaming binge* is a relatively new phenomenon that's been created by the move from live or recorded media to a content feed that is offered on an as-desired basis. This was as much a game changer as the VCR and the DVR, in that it supercharged the concept of convenience to an entirely new level. Now you could enjoy a movie or favorite TV show right away without driving to your local Blockbuster video store (remember them?) or having to order a DVD by mail like in the earlier days of Netflix. Those older methods seem so far away now — almost a different era. But it was not all that long ago. In the realm of technology, time indeed moves at light speed.

WARNING

Enter streaming — a feed of digital bits and bytes that make up your favorite video entertainment. What could be bad about this? Well, I hate to be a party pooper, but the ability *to get what you want, as much as you want, and when you want it* is not always a great thing. And if this sounds familiar, it is because it describes the same dynamics that helps make the Internet so desirable and at times addictive.

As you find out in the following sections, streaming offers endless choices, along with ease of access and stimulating content. It also lacks something that most other forms of communication and entertainment have — *boundaries.* Boundaries remind you of natural limits and where a story begins and ends. They serve as markers to help you set limits on your consumption, and the fewer the boundaries, the more likely you'll just keep watching. Streaming content developers count on this fact. They want you to keep watching (keep in mind that the profit engine of the Internet is eyes-on-screen). This is the reason why the default option on YouTube, Netflix, Amazon Prime, and others is the *autoplay* feature that automatically takes you to the next episode. It will even lead you to a similar show if you've finished the one you were watching and then start it automatically. The other change is that many TV shows and series are released all at once, so you have an additional impetus to binge, although some shows are still released weekly. This is equivalent to opening a bag of potato chips and eating them until you hit the bottom of the bag. After all, who can just eat one?

Understanding the allure of endless choice

So, what is it about *endless choice* that is so darn compelling? On the surface, having endless choice seems like a good thing. After all, what could be bad about having tons of choices? Well, let's look at a brief analogy. Recently, I went into a convenience store to get a cup of coffee. I was in a rush, so all I wanted to do was run in and run out quickly. Well, no such luck. First, I had to decide on what size cup and choose from *five* different options, followed by *eight* types of coffee; then there were the creamers, *twelve* in all; then, the *six* types of sweeteners. Well, it may be hard to believe, but all those valuable choices produced nearly 3,000 possible combinations and options. All I wanted was a cup of coffee, not a life decision, yet that is where I ended up.

You assume that *more is always better,* that more options mean more opportunity for satisfaction. But too much choice can produce complexity and stress. This is exactly why I like shopping in small grocery stores — fewer options to choose from. This leads to a simpler experience and thus less time wasted, and less stress.

The same is true for the endless opportunity of watching 600 channels of cable, and if you add in the two dozen streaming sites and services, you have even more choices. Your brain doesn't really need this much choice, and when you have too much online content to choose from, it can add to what I call *Tech Stress Syndrome* (see Chapters 10 and 12). Sometimes, *more* is *not* necessarily better. I am reminded of the four or five TV stations that we had when I was a kid; I recall often saying to my parents that there was nothing on TV — can you even imagine hearing that today?

REMEMBER

Endless choice compels your brain to think there is always a different potential option. Options are good sometimes, but when it comes to digital content, you can easily become mired in the lure of promised entertaining experiences to keep you satisfied, but this satisfaction comes at a price — your time and attention.

Finding the power of instant access

Faster is better, and I would be lying to you if I said I do not appreciate faster Internet access. There is nothing fun about waiting for content to load while staring at a buffering screen. I recently had the opportunity to graduate from our DSL connection at 3 to 5 megabits per second (mbps) — yes, that's right, 3 to 5 mbps. That is barely faster than dial-up! I was elated, but at the same time I knew that faster speeds (now at 50 mbps) meant a faster life, more choices, and potentially more stress. These increases in speed are a double-edged sword, as there is always a dark side to all tech advances.

The power of instant access is complicated by the fact that instant gratification capitalizes on a basic addiction and pharmacology principle. The shorter the period between when you click something online and when it shows up on your screen (and arouses your nervous system through elevation of dopamine), the more addictive that behavior is. This is the same with substance use. Drugs taken intravenously (the fastest route of administration) have greater addictive potential than if taken via another way, such as snorting or smoking. Faster devices, faster processors, and faster Internet all equal potentially greater risk of Internet and technology addiction; even if you aren't addicted, it can contribute to a greater tendency to abuse or overuse your screen technology.

REMEMBER

There is a benefit in delaying your gratification through instant access; delay gives you the ability to think through your choices and make decisions based on executive skills and reasoning that are often inaccessible when engaging in an intoxicating and immediately gratifying experience. There is a lot of power and distraction potential from the instant access that streaming provides; the ability to satisfy your instant desires without interruption makes endless digital content easy to overuse and at times addictive.

Recognizing the pitfalls of effortless starting

Effortless starting (autoplay) is the ability to passively allow your television or online screen app to deliver the next episode, video, or content automatically, without you *doing anything*. This passive approach, as I mention earlier in this chapter, is how streaming content providers run their systems; unless you purposely opt out of this feature, you'll be effortlessly fed content until you either turn it off or walk away. But how many times have you sat in front of your screen and allowed the next video or episode to come on and swore that this would be the last one? I've certainly done this. *The problem here is that addictive behavior is facilitated by doing nothing.* Effortless addiction is perhaps the most insidious because you never even have to move off your couch.

The organizing part of your brain doesn't really like endless or unfinished things, or rather, it doesn't like to leave them endless. This is in part because you're neurologically designed to finish and complete perceptual and cognitive tasks. This concept of completion is sometimes called the Zeigarnik effect and is borrowed from the perceptual and cognitive science of how you perceive the world around you. It's quite simple. If you draw a series of numbers that reflect a pattern on a whiteboard, your brain will naturally attempt to complete the pattern; this is also true if you draw a near-circle on the board. Even if the lines do not meet, if they are close enough together, the perception will be of a circle. The reason for this principle of *closure* and *completion* comes from your brain being designed to organize and categorize as efficiently as possible — putting things in order allows expediency and predictability for living in the world. This trait has helped humans survive and allows you to create shortcuts for managing your environment.

When digital media is endless, however, your brain still tries to finish it. You want to get to the completion — to finish that process. The problem is that with a seven-season TV show, it is going to take a lot of time to get there. The Internet is never done, but your brain nevertheless tries hard to finish it.

All *boundaries* in dealing with the Internet and anything streamed must come from your *conscious intention* and behavior, to go against the natural ease of access of your streaming screen.

Unpacking user experience engineering

This is the million-dollar question: How much social and behavioral engineering goes into Internet entertainment (this can include news, social media, movies, TV, shopping, video games, and more)? The answer is plenty. The Internet medium as a digital modality is designed and managed using specific algorithms that are unique to your online history, demographics, and use patterns. Your *user experience* is sculpted to sell you content those producers and providers have to offer, or if they do not offer it, they will sell your name to someone who does. The technical term for this is *user experience engineering*, which is a fancy way of saying personalized manipulation. This is hardly a nefarious endeavor. These people are not out to get you, but they *are* out to get your time and attention and make money. The question you must ask is "How much time do I want to give?"

So, what kind of brain science is going on here? Well, a lot. Behavioral psychology and neuroscience (along with social psychology and consumer science) all have input into understanding human screen behavior. Video game developers will go to elaborate lengths to test sounds, music, A–B comparison of colors, and reward contingencies to arrive at a perfect formula that will enhance your user experience. But what exactly does *enhance the user experience* mean?

REMEMBER

An *enhanced user experience* is code for an experience that produces a long use period. Whether this is a game, a TV series, or other digital media content, the idea is to get your eyes on board and to keep them there if possible. Unfortunately, your body (and life) tends to be where your eyes and attention are, so this all adds up to a heck of a lot of time on your screens.

All reinforcement and reward contingencies used by providers are well researched and based on both animal and human studies. There is over a hundred years' worth of data on how to increase specific behaviors and how to reward people in a manner that increases the likelihood of maintaining that behavior. This is science that has been experimentally proven, and all this data (as well as new research) is available to the content producers, developers, and service providers that bring you the content and experiences that you enjoy. The question is, do you *really* love it, or do you *love* it because of the *way* it is given to you?

TECHNICAL STUFF

The concepts of boundaries, stimulating content, synergistic amplification of content, reward latency, variable reinforcement, interactivity, and endless content all represent well-researched topics in behavioral and consumer science. The fields of cognitive brain science and cognitive psychology have devoted a lot of research to how to manage attention and how people process information. I recently wrote a chapter in a book titled *Human Capacity in the Attention Economy,* edited by Dr. Sean Lane and Dr. Paul Atchley. The focus of this book is on attention and the psychological, neurological, and economic factors that operate in the attention economy. We have now moved from the *information age* to the *attention age,* and whoever gets to manage and control your attention has tremendous power.

Seeing the influence of social media

The integration of streaming, social media, and social influence on consumer behavior is a potent new force. The ability to utilize people (who are not celebrities) to help sell and influence purchases marks a new dawn in the world of marketing as well as entertainment. No one foresaw the use of social media and the way it has evolved over the last decade; short videos that stream on YouTube, Instagram, and TikTok capture people's attention like flies on flypaper. The use of social capital and social influence on marketing and sales provides even greater power to the Internet medium to impact people's lives and decisions. Even the news uses gamification and social influencing to keep your eyes onscreen. It's impossible to insulate yourself from this process if you use the Internet, and frankly, nearly everybody who has Internet access is affected and influenced in some way. You use the Internet not only because of what you can *do* online, but also because of what it *does to you.*

REMEMBER

Television, and to a large extent the Internet, needs your eyes onscreen to fund its existence. Anything you see streamed online or on TV (and the lines between these media have now blurred) is *funded* with your attention. You always pay with your time, and it's important to ask yourself how valuable your time is.

TEACHING YOU THAT YOU MUST SEE WHAT'S ON NEXT

Autoplay is almost as important an invention for YouTube, Netflix, Amazon Prime, Hulu, and most streaming services as the *Like* button was for Facebook. What do I mean by this? Well, consider this example: It's Thursday evening, you've had a heck of a week so far at work, and you are looking forward to sitting on the couch to start to watch the newest season of your favorite show that was released the night before. It's 9 p.m. and you turn on your smart TV and choose one of the streaming services that carries, and more than likely produced, the show. There is a whole new season of ten episodes, and you are going to watch the season opener and then get ready for bed, as you have one more weekday to get through, and you need to be at work early for a meeting.

The show was amazing, everything you hoped for. The ending was so suspenseful (and left you wanting more) that you let the credits roll while you stretched, and the next thing you know, there is a little circle that begins to turn at the bottom-right corner of your screen, telling you to hit cancel if you do not want the next episode to play in the next ten seconds. Ten, nine, eight. . .*should I just watch one more? I can probably watch one more and still get to bed at a reasonable hour.* Seven, six, five. . .*okay, what's one more episode going to do anyway? It's not like I'm going to binge the whole season tonight.*

The next thing you know, it's 4 a.m. and you have just completed seven of the ten new episodes — and you have to be up in three hours for work. What happened? All you know is that you sat down to watch a show, and instead, you watched *seven.* What happened was *autoplay.* This is a very simple but powerfully clever method of you doing nothing and it getting the content producer and service provider something very valuable — your time and attention — which translates ultimately into money. You were a victim of what most of us have experienced, probably more than once. By doing nothing, you got to *passively consume* video content; this is because your energy would have to be spent to *not* do something, as opposed to *doing* something. It is a perfect use of Newton's law — that is, an object at rest (and lying on the couch counts) tends to stay at rest unless acted upon by a force. The same also applies to an object in motion. The problem is that at 2 a.m., there are not a lot of forces hanging around to help.

Looking at Other Problems of Watching TV All the Time

The problem with television is that in some ways, it is too good. After all, where else can you tune in 24 hours a day and be pleasantly entertained or distracted — except the Internet? The issue is further complicated by the fact that the Internet and television have finally merged to a large extent. Thanks in part to streaming shows and services (many competing with cable programming), the ability to find, watch, and binge TV today is perhaps greater than ever.

In the earlier days of television (I grew up in the 1960s), there was not much to watch, and on most stations, broadcasting ended at 1 or 2 a.m. There was no Wi-Fi, no Internet streaming, no cable — just a few television networks transmitting through the air without much competition for viewers' attention. The problem with watching TV today is that it is *always on*, with hundreds of choices, and if you add streaming, there are even more choices. The ease of access of TV integrated with the Internet is that you now have television on *steroids.*

I talk about the ease of binge watching earlier in this chapter. In the following sections, I go over other problems and pitfalls of watching TV all the time.

Intensity is addictive

When it comes to television, *bad news* is compelling. Not that you like bad news, but there is something about scary or bad news that keeps you watching. It's likely that this is a primitive self-protection mechanism in that you need to know whether there is any potential danger; the problem is that it compels you to feel that the world is unsafe and dangerous, when in fact, it is safer today than it was 100 years ago. However, because you are constantly immersed in instant access to all the bad news happening all over the world, you feel as if danger is lurking around every corner. This is not to say that bad things do not happen, but before television, and especially before Internet-linked, cable, satellite, and streaming television, it took a while to find out.

Near-instant access means that you will hear and see everything that happens everywhere in the world, almost in real time. This is not bad or good, but it is certainly a potent force in impacting your emotions and consciousness. Being informed is important. Being saturated with bad news that leaves you feeling stressed, overwhelmed, and scared is *not.*

There must be a balance in the Internet attention age. Digital screen technologies are wonderful and help you do many great things, but they also add a layer of stress (along with eating up your time). Many times, you can get caught up in the amazing things you can see, do, and hear online, but less often do you address how much time and energy all this *convenience* costs you.

TV acts as your social companion

We're lonely. In fact, we're lonelier today than we were 50 years ago. We have more square footage in our homes than ever, but at the same time, fewer close, personal friendships than ever. We consume a lot of material goods and intoxicating Internet content, but we also find that it contributes to feelings of further isolation. We often view television as a companion, with the soft drone of whatever is on as background noise, acting as a backdrop to our lives. It's not that TV is bad, but just like the Internet (which has essentially merged with TV), it's almost too good, and like chocolate ice cream, too much of a good thing can hurt you.

Television has become a place to hang out with your shows, movies, and content (all provided with autoplay, as I describe earlier in this chapter). You can be passively entertained for hours upon end without moving a finger. But this can really become too much. I recall how appealing the thought of watching several episodes in one night was, but when I gave in to this temptation, it was usually at the expense of something else — namely sleep or a task waiting for my attention. The problem here is that pleasurable behavior (which also happens to be addictive) is provided and consumed in such completely a passive manner — essentially, you do nothing except talk into your remote. This is the path of least resistance, and changing this pattern requires active choice and movement.

One form of screen use is almost as good as another

When someone stops heavy Internet and video game use, it's not unusual for them to switch to watching huge amounts of TV. Because television is now integrated with the Internet, it is easily overused and at times can become an addiction itself. The ability to binge on TV with streaming capability makes it very easy to lose track of time. The word *stream* really describes the endless flow of content without interruption. The ability to choose from hundreds of options, and to have instant access, makes stream-based television a whole new game. Many patients I've treated have developed television addictions once they no longer had free access to the Internet or video games, and so they were essentially swapping addictions. It is important to try not to switch from compulsive Internet or video game use to excessive television viewing, especially when the line between TV and the Internet has essentially disappeared.

TECHNICAL STUFF

The average daily use of the Internet has now surpassed television usage, but it's hard to tell where all this will lead when distinguishing the Internet from television is becoming steadily more difficult.

It's a Choice: Screening the Stream

Because you have limited time to spend on your screens, including TV, how do you know what to watch and how much is too much? The idea really comes down to *choices*. You make choices every day in terms of how much time you spend on everything you do; the trick with screen technologies is to apply this principle of conscious choice to the seemingly harmless aspects of staring at a screen. Whether TV, the Internet, or a video game, it all eats up your time and attention, and often in a way in which you cannot appreciate what you've been doing for all those hours (and what you neglected in your life).

You must make choices, and because you cannot do everything, you need to make sure that you see your TV and Internet use as an *expendable* activity, not a *necessary* one. I am not talking about the work or school activities you do online or the show you like to watch. I am talking about all the hours you spend endlessly channel surfing or Internet scrolling, hoping to see something that gives you some hit of pleasure. Because you often cannot appreciate the passage of time when staring at a screen (see Chapter 1), you must make those decisions *before* you start watching.

TIP

Set limits on how much time you want to watch and stick to those limits. Think about the other valued activities you have in your life and how those hours of watching will impact your sleep, health, relationships, work, or academic performance. Your technology and screens can be wonderful additions to your life if you apply your humanity to them; left to their own devices (pun intended), your screens will slowly encroach on your essential humanity, and you will lose what is truly important. Flip to Part 4 for more tips on living a balanced life with the Internet and technology.

REMEMBER

Whether it is Internet-enabled TV, where you can stream almost anything, or old-fashioned broadcast TV, your central power is from you deciding when to turn it *on* and when to turn it *off*. You cannot wait for an external reminder to limit your use or do something else, because that reminder may never come. You must remember that all these wonderful technologies are here to *serve you*, and if you are not careful, you will end up *serving them* instead.

Chapter **7**

Exploring Video Games and Video Game Addiction

Video games — the subject of this chapter — are a part of everyday living. Now considered a sport, with championships, televised competitions, and professional teams and players — video gaming has indeed gone mainstream. Estimates are that 90 to 95 percent of boys and perhaps as many as 60 percent of girls play video games, and the age of play is not limited to children and teens. Many young adults, and adults of all ages, play as well, and most people who do so have few issues, if any, with abuse or addiction. In other words, most people use video games as they were intended, as a hobby or pastime that is fun and stimulating.

Prevalence of video game addiction in teenagers varies around the world and may be as high as 50 percent in Korea and China, in part because of differing measures and definitions for the addiction. Overall, it appears that video games are in fact addictive for the small percentage of users who cannot seem to limit their use. For those individuals, video gaming is no longer just a game, but rather functions like a digital drug. For the approximately 1 to 5 percent of gamers in the United States who cannot stop or say no, video gaming has become an obsession. Many more gamers may overuse video games as well, although not to a level that meets the criteria for an addiction. For millions of video game addicts, normal life balance is often lost and replaced with preoccupation and excessive use of video games. If they aren't gaming, they are *thinking* about gaming; if they aren't thinking about gaming, they are watching a live or recorded stream of someone gaming on YouTube or Twitch.

For the patients who come to me (often at the suggestion of their families) because they are gaming so much, their lives have essentially become unmanageable. They are unable to stick to moderate use and are not able to walk away from the game. I've seen many patients over the last 25 years whose lives were so derailed from their video game addiction that many of the reasons why family members contacted me about their loved ones were exactly the same reasons noted in dealing with *any* other addiction. People wanted their child or family member back. They did not want the addiction to control their loved one's life to the point that nothing else seemed to matter. Perhaps the most impactful part of addicted video gaming is that at the end of the day, battle, or tournament, *all you have is nothing.* Years of video gaming leave you with little that is transferable to real-time living; not that fun is a bad thing, but fun all the time may be a negative.

Wondering How a Video Game Can Be Bad for You

When studying addiction medicine, my initial training and experience was with people who abused substances — mainly drugs, alcohol, or smoking. Occasionally, I would treat someone with what is called a *behavioral* or *process* addiction, such as gambling. With such addictions, you do not physically consume anything. With gambling addiction, nothing is ingested into your body, yet you still can become addicted, with many of the same symptoms and life impacts that occur with drugs and alcohol.

This is because addiction is not simply about the substance or behavior, but rather about the total *pattern of addiction* involving multiple brain centers and numerous psychological variables; these changes produce a pattern of living that is reduced to the point that the only thing fully attended to is the drug, alcohol, gambling, or, in this case, video games — this is covered in more detail in Chapter 2, where I review the biology of addiction.

WARNING

In 2018, a 13-year-old girl in Mississippi died after being shot by her 9-year-old brother over a video game dispute; it seems the boy grabbed a gun one afternoon after his sister would not give up the controller. Although this level of violence is by no means a typical scenario, it does speak to the power and potency of the drive to play.

A BRIEF HISTORY OF VIDEO GAMES

If you're a bit older, you may recall the early days of video games. Not recognizable by today's standards, they were essentially primitive when compared to today's sophisticated multi-user console and PC-based games, which have stunning graphics, realistic images, and high-definition sound. Early video games simply involved a few pixels making simple movements on the screen. Titles like Pong represent those very early days, but I recall the excitement I had in the early 1970s when we got our first basic gaming console and we hooked it up to the antenna terminals (do you remember TV antennas?) on the back of our 150-pound cathode-ray color TV. It seemed miraculous to be able to manipulate an image on the screen by using a hand motion, and as basic as it was, it was still captivating. I think it was this newfound sense of power, control, and interactivity that was so appealing. Prior to this, TV had always been a one-way experience, and we were passive recipients of the images on the screen; now we were controlling the images and competing against another person in a game.

Greater sophistication soon followed in games like Asteroids and Tank, with each iteration becoming more technically complex, challenging, and fun. With some games today, you *cannot* discern a game image from a real image — that is how good the graphics have become.

Video games are now a major industry both in the United States and around the world, with the U.S. video gaming industry bringing in well over 100 billion dollars annually. This is no longer a small industry, and because of this, there is a strong incentive on the part of the industry to keep that revenue flowing. My experience in dealing with the video game industry is that there is a degree of corporate denial regarding the percentage of video game users who become hopelessly addicted to gaming, and thus experience significant negative life and health consequences. When I appeared on an episode of HBO's *Real Sports with Bryant Gumbel* in 2019, the lobbyist for the video game industry essentially denied that video games could be addictive, despite considerable scientific evidence to the contrary.

There is a long history of various *pleasure industries* denying the addictive nature of what they sell, only to eventually be forced to acknowledge the truth and ultimately participate in socially responsible education of their users and families. The alcoholic beverage industry, tobacco industry, and gambling industry have all eventually admitted to the addictive potential of what they sell and acknowledged that some people do develop a problem that will need treatment. The video game industry has not yet arrived at this level of responsible admission, despite the World Health Organization (WHO) declaring video game addiction an official medical disorder in 2019.

(continued)

(continued)

The irony of the video game industry's denial is that essentially all addictions share the same underlying neurobiology, and for the most part the brain cannot differentiate the source of its intoxication. In other words, it doesn't really matter what you become addicted to; once addicted, you must deal comprehensively with the addiction. Addiction is not simply about the substance or behavior that changes mood and elevates dopamine. It is an entire process of behavioral effects, including changes in memory, reward, self-care, motivation, emotion, socialization, and goal-directed behavior. The actual substance or behavior that starts the pattern is just the tip of the addiction iceberg.

In full disclosure, my gaming expertise and prowess never extended much beyond Asteroids, and I was even bad at that, so it was never all that much fun for me in the long run. I do, however, get the idea of how intoxicating *building* or *controlling a new world* in a game can be, especially when compared to building a model airplane.

REMEMBER

It is *not* the video game that is bad for you; rather, the *addiction syndrome* that is developed by the addictive use of the game is the real problem. The video game is simply the catalyst that starts the addiction cycle, and like a car engine, once started, the cycle keeps going as long as you keep fueling it. Video gaming is the *fuel* that keeps a video game addiction going.

People think I am anti-gaming. That is inaccurate. I am anti-addiction, and if video gaming is contributing to an overall pattern of addiction, then it is a problem.

Understanding What Makes Video Games So Addictive

What makes video games so addictive? This is the million-dollar question, and some aspects of this question are also covered in Chapters 1, 2, and 4. Video games are addictive for a variety of reasons, as you find out in the following sections.

Providing stimulation and variable rewards

A video game is designed to be stimulating and very rewarding. The game designers are trying to create a lush and exciting gamescape that holds your attention, stimulates and challenges you, all while providing variable and intermittent rewards. These rewards must be unpredictable, because if you know what to expect in the game, you will quickly lose interest.

Game designers and behavioral (experience) engineers use well-established behavioral and neuroscience principles to structure the game in a way that keeps you not only riveted to the screen, but also wanting more. This is not an easy task, because the game must be designed in a way that is playable by most users, but that will also offer increasing challenges, allowing a user to level up within the game. This is very important, because it also applies to the thrill of mastery and the elevation of your comparative social stature within the game. As you move up the ranks, so to speak, you gain more power and prestige; all the while, your brain is being neurochemically conditioned to respond to the variable rewards with dopamine spikes and a lot of anticipated elevations of dopamine from possible future rewards as well.

REMEMBER

Gamers keep playing because the expectation of possibly getting a reward within the game is very rewarding itself and is followed by a twofold increase in dopamine compared to actual hits you receive in the game. What this means is that the *maybe* factor that is inherent in the game is a powerful reward in and of itself. *The possibility of getting a reward is rewarding,* so you keep trying over and over. All the while, your brain's reward center is lighting up like a Christmas tree; this is similar to a gambler wanting to keep gambling regardless of whether they are winning or losing. (Flip to Chapter 2 for more on the biology of addiction.)

You don't see or feel any of this. The process is completely automatic and unconscious. All you know is that you're having fun and that you love the game and cannot wait to play it again. This is how an addiction takes root. For some people, this cycle becomes chronic, and they start to develop brain and behavioral changes. Through the process of up-regulation and down-regulation, your brain acclimates to increased dopamine levels, and changes occur in how much dopamine you release from other previously pleasurable activities. This is called *reward deficiency syndrome,* and it looks a lot like depression (see Chapter 2). Things just don't feel as much fun, especially when compared to video gaming. The scope of your life starts to narrow, as things that were previously important or pleasurable no longer feel all that rewarding.

While all these changes are occurring, there may also be changes in your academic performance — school may not seem as important, and your grades might start to fail. Extracurricular activities or a job may be of less interest, while sleep and physical activity may also be reduced. You may be spending less time with friends and family, and become more isolated in your room, house, or apartment, playing your games. Although it may feel fulfilling, your life will be narrowing. As the cycle progresses, you might take less care of yourself physically, with less exercise, bad eating habits, and poor hygiene and self-care. Not everyone follows this progression, but in my experience, most video gaming addicts go through at least some of these changes.

This process is slow and gradual, taking many months or even years, but it typically progresses to a point where there is significant life impact, and then it cannot be ignored. Well, the addict can ignore the consequences of their growing addiction — in the form of denial. The addict's denial allows them to not see themself as having a problem, even though everyone around them does. It is important to remember that this is not simply borne out of oppositionality or willfulness, but rather out of their blindness to an addiction. They are unaware of the gradual intoxicating shifts and changes that have occurred over time, and they may insist that there is nothing wrong and that they do not want to give up their game.

Feeling a strong sense of identity and competence

Everyone wants to feel competent at something. If you're like most people, you derive some self-esteem from how well you perform at various activities and tasks. The same is true for video gaming. Most gamers feel good about their game performance and levels of mastery in a particular game or games. They take pride in the skills they develop, their ranking and ratings, leadership, and how they level up and compare with other players within the game.

To some extent, your level of self-esteem becomes based on how you compare to others and your self-appraisal of how you're doing at a particular game task, and you also respond to how you believe others view you. Notice that I said how *you believe*, as you really don't *know* how others see you, but you may often judge yourself on your beliefs based on those assumptions.

Because everyone wants to feel competent (and although comparison is said to be the thief of happiness), you're always comparing yourself to others. In video gaming this is amplified by the fact that everyone's game performance is on display for everyone else to see, and you can always examine your scores and ranking based on these comparisons. If you're doing well (obviously, this is somewhat subjective), then you may be deriving some, or much, of your identity and self-esteem from the game and within the social context of the gaming community.

Depending on video games for identity

If you are a family member or friend of a gaming addict, then you must be able to appreciate how much a video game means to the addict. It is not just a game. Often video gaming becomes the central part of a video game addict's life, and when this source of fun and self-esteem (as well as dopamine) is threatened, it can become pretty upsetting to the video game addict. Their entire identity becomes wrapped up in the game and gaming community such that there is little opportunity for them to find meaning and positive self-experiences without the game, at least initially. You can certainly understand why the addict may be reluctant to give up what they know makes them feel good, despite the negative problems it creates in their life.

REMEMBER

Part of the problem is that to a family member or friend of a gamer, the video game is, well, a game. The word *game* can be misleading, and it can convey a lack of importance to loved ones of an addict. To an addict of any kind, including drugs, alcohol, gambling, and even sex, their drug of choice is not simply a form of recreation — it is very important to them. It may be something that a non-addict would see or do as a recreational activity, but to an addict it means a lot more. To an addict, not only are their identity and dopamine levels dependent on the game, but they typically spend so much time on the game that they literally do not know what to do with themself unless they are gaming, watching a game, reading about gaming on Reddit or elsewhere, or simply thinking about gaming.

Everyone likes to feel that they are good at something, so when you suggest to a gaming addict that they should give up their gaming, at least initially, you should keep in mind that there is little else to replace it. Their ability to derive satisfaction from real-time living is largely absent. They are so dependent on the game for their identity and personal gratification that they can't imagine living without it, and they have allowed their social and familial relationships to atrophy to a point that they no longer know how to feel connection and intimacy. By the time I become involved for a consultation or treatment, there have usually been numerous failed attempts at various treatments, as well as attempts by parents or loved ones to remove devices or limit game use. Most of the time their efforts fail — not for lack of desire on the family's part, but usually because it becomes logistically difficult to prohibit gaming without some focused and consistent changes on the part of the addict.

REMEMBER

Many times, parents or family members have provided counseling or psychotherapy, including medications, for the addict. In my experience, most of these treatment attempts fail because they don't target the primary problem: the addiction. It's not that there are no other psychiatric issues, as there often are, but you must keep in mind that an addiction is essentially an attempt to manage many symptoms, *and then the addiction itself becomes a key problem*, thus increasing other psychological issues. You must address the addictive pattern and then begin to unwind all the other issues throughout the treatment process. Depression, social anxiety, ADHD, and other psychiatric and behavioral issues often co-occur with a video game addiction, but care must be taken to target the dysfunctional addiction pattern; otherwise, other interventions are likely to be ineffective and the addiction will continue.

Switching the addict's sense of competence to other sources

There is often very little else in life that a video game addict would like to do when compared to playing a video game. It is not that there is no other fun or meaningful activities, but few are as easily *accessible, desirable,* and by now, as *familiar* to the addict. They have long forgotten how to have fun in other ways, so to simply make suggestions of what they could do instead of gaming will likely be met with resistance at best, and anger at worst.

How do you get the addict to switch to other sources of self-gratification and identity, along with more constructive sources of dopamine? The difficulty here is all about timing. Unless you have a period of detox where the behavioral patterns of the addict are disrupted, you'll have little success in getting them to try anything else. The reason for this is simple. Why should they do anything else if their gaming is making them feel so good? More importantly, they cannot *perceive* that there might be something from the past, or something they have never done, that could be exciting or stimulating. The only thing they want to do is game.

TIP

The trick is to begin to reintroduce real-time activities while the addict is in a detox period (and after the initial withdrawal period), when they are no longer gaming. In Chapters 11 and 13, I discuss whether the gaming addict can return to moderate use or may have to abstain from gaming altogether.

REMEMBER

The transition between gaming life and real-time life is often a *gradual* process, but my experience suggests that if you can get the addict past that initial, difficult withdrawal and detox period, then they can eventually come to find other life activities gratifying. This process can take months to accomplish and is often fraught with ups and downs, potential relapse, and frustration for both the addict and their loved ones.

Using complex skills in an exciting way

Video gaming is a complex skill. Although it may have the reputation of being a *slacker activity* to some, it is a sophisticated and challenging task that requires intelligence and problem-solving skills. In my 25 years of treating video game addicts, most have seemed to possess above-average intelligence.

Video games are complicated. Although the broad category of gaming encompasses many different types of games and therefore many skill sets, they all require attention, concentration, focus, cognitive flexibility, fine motor skills, and good problem-solving ability; these are often the traits we associate with general intelligence, and these skills, if applied elsewhere, would serve the video game addict well. The problem is that the gaming addict often has *little motivation* to do much else besides playing the game. The goal is to redirect these talents in other directions that will provide some of the challenge of the game, along with the rewards that come from more balanced living. This is no simple feat.

The game offers many things to the addict. Mainly, gaming is a chance to use their abilities doing something that is entertaining, challenging, creative, and exciting. Excitement is key here. Video gaming allows a user to feel an enhanced sense of *mastery* and *accomplishment* within the game world. Sometimes this involves beating a personal best, learning a new game, or other times, it involves beating other players or other teams. Either way, the gaming addict is often getting positive reinforcement from their gameplay, both extrinsically from others and intrinsically in term of self-appraisal — and, of course, continuous hits of dopamine seal the deal.

REMEMBER

Take care not to assume that gaming is simply about a game. Video games are often highly technical and sophisticated puzzles to be solved, and the gaming addict is often immersed in a way that provides a great degree of the meaning and excitement in their lives.

Becoming part of something bigger than yourself

Everyone wants to be part of something bigger than themself. Video gaming offers the opportunity to join a group of people who all share the same interest, language, and passion. Every day, people are attracted to different clubs, fraternal organizations, and activities with like-minded people. It's no wonder that video gamers and those who become addicted to video games share the same desire to be part of some larger community. There is a sense of power and belonging that you get from being a member of a gaming guild, clan, or team with a special mission, goal, or purpose that you sign onto.

REMEMBER

When you share goals and activities, this creates a common purpose and connects you with others. The video game community offers this connection, and because many games are played through interactive and competitive groups or MMORPGs (massively multiplayer online role-playing games), these groups essentially become social groups. Obviously, much of this connection is virtual. Shared video feeds and voice discussion during and after gaming via Discord or other methods allow for social connection; social networks are built among the teams, groups, guilds, and so on, and this, in essence, becomes a community. Although you might argue that this is not a real social connection, I believe it is of a sort; the problem, however, is that for the gaming addict, *all* their social connections are built around gamers and gaming, allowing for little else.

What holds all groups of people together is a combination of shared goals, group norms, and rules. The gaming community has these goals, norms, and rules just like any other community. It shares an Internet gaming passion and carries with it a language that is specific to the world of gaming. As I note earlier in this chapter, the gaming community in the United States is enormous, with most people using their games with few issues or problems. But for those who develop an addiction, the insulated nature of the gaming community creates a place where they can hide from the realities of the negative impact that their addiction has on their life.

Pride is very motivating. Feeling that you're part of something that is *bigger* than yourself and where you have a responsibility to others in the group creates a sense of belonging. I'm thinking of the pride and shared goals in a military organization or a club. The idea that you're *part of something* pushes you to face more adversity and take on more challenges. All the while, you're connected to, and surrounded by, other like-minded people. If you're part of a cohesive gaming group, you have a shared mission, and that feels motivating and important. Competition builds pride, and in the gaming world, there is a lot of competition.

With video gaming, like every group or specialty interest, a language evolves that is shared among the members of the group; this has the effect of creating a bond among the users who are inside the gaming world or within a respective game. A shared language builds cohesiveness and creates boundaries around the group that outsiders can't penetrate. By separating their gaming group, clan, team, or guild from others, through their own language and rules, users further enhance the uniqueness and identity of their experience. Addiction isn't caused by this separation, but rather, it creates the sense that everyone is part of the group and where gaming is the norm.

Elevating social status and finding respect in the gaming community

How do you raise your social status in any community? Social status is acquired through paying your dues, coming up through the ranks, and proving yourself through ordeals. This is how a group gets to see that you are one of them; when you go through what others who came before you went through, that shared camaraderie creates a sense of belonging and trust. This is how a fraternity works and is why going through the challenging activities while pledging secures the trust and respect of those who came before you. This rite of passage binds the past, present, and future together, and in so doing, creates a powerful allegiance and connection — you trust those who do what you did.

This practice is as old as human society, and the video game community is no different. Getting through a quest, battle, or siege as a group bonds you together, and that bonding becomes the basis for strong loyalty. Many of my patients were *so loyal* to their gaming group that they felt significant guilt and remorse for disappointing them by leaving the game.

TECHNICAL STUFF

Many of the gamers I've treated had long histories of low self-esteem and a feeling that they were inferior to or different from their peers. In many cases they suffered from ADHD or social anxiety. In some cases they did not fit in with more traditional groups and roles among their peers. However, once they found video gaming, and especially multi-user gaming, they felt at home. They liked being part of a group (and who wouldn't?), and sometimes these groups also enjoyed playing Dungeons and Dragons (D&D), Magic: The Gathering, Pokémon, or other games. Obviously, COVID has impacted some of these in-person activities. Without making too broad a generalization, I've found that many of these more internalizing games and activities are attractive to those kids who later went on to develop a video game addiction. That does not mean that if you enjoy these activities, you will develop a video game addiction, but I've found that if a video game was not available, these other activities were often desirable to many of the gaming addicts I've treated.

Connecting on a shared challenge

As I note earlier in this chapter, we all want to belong to something bigger than ourselves. Multi-user gaming is perhaps the most common form of video gaming currently. It is played via the Internet, and users are connected through teams, groups, and competitions. The Internet allows interactive groups to play on PC-based and console games, and gives players the opportunity to connect and play as if they are physically together. They can view representations of themselves as icons and avatars on the gameplay screen, almost as if they are inside the video game, and they can also talk through an audio connection with headsets via Discord or another app.

This type of gaming is very appealing in that it allows the user to have a social and shared experience while being in two distinct locations. The Internet's ability to connect people and places together offers a perfect solution for video gaming in that it creates a group experience from a solitary one. People can have group goals and experiences and are able to enjoy social interaction, even when separated by great distances. Physical boundaries and distance are meaningless with online gaming.

Part of the difficulty in managing an online video game addiction is that it meets many of the social needs of the user, and when limiting or leaving the game, they have even fewer options for social connection. This issue has become even more amplified during the COVID pandemic, where interaction via the Internet may be their only option for socializing.

Let's face it: We all want and need to belong. We are social animals, and we have a need to connect socially and physically, as do many animals. We are hardwired to have social connections, and online connections that users experience when gaming can fulfill these needs. As noted earlier, we also want to be part of a group and to feel that we are members of something *bigger than ourselves.* Having this sense of belonging to a group is very appealing. The issue with video game addiction is that gaming tends to become the only source for such connections and can limit your motivation to seek connection in other ways.

Changing the rules as soon as you think you know them

How do you master something that can never really be mastered? Video game designers are experts at developing rules and algorithms that create and maintain a challenging gaming environment. The key word here is *maintaining,* because to maintain a challenging game, rules and game formulas must be changed and upgraded on a regular basis to sustain the challenge. There is no easy or immediate mastery of a game because the road map to that mastery changes all the time — just enough to give you satisfaction, but not so much that you become frustrated at never being able to achieve competence. It is a Goldilocks formula of creating just the right amount of fun and challenge, as well as dynamic change. If it is too easy, you will not feel challenged — too hard and your frustration might lead you to play less or, even worse, abandon the game.

The problem for the video game addict is that there is no ability to finish the game easily, and perpetually evolving challenges help facilitate an addictive experience — namely, because there is always another new experience to be had. Imagine a roller coaster that changes each time you ride it; a well-designed video game is a lot like this. I remember going to Disney World in Florida years ago. I recall that the thrill of one vertical and horizontal high-speed ride was that its ride pattern was computer generated randomly, meaning that the ride changed each time you rode it. How terrifying (for me, anyway), and how thrilling and captivating at the same time. *Variability* and *unpredictability* are appealing drugs of choice.

How can you ever hope to master a moving target? You cannot — and that is why game developers keep it moving. The object is to secure a player for several years of gameplay, and to do this, they must keep the game a moving target.

Games are often designed with increasing challenges and updated versions offering new features and opportunities. You are continually improving your game, using new strategies, and solving new dilemmas within the game, and then a new version comes out! This is incredibly compelling and a lot of fun, but it also gives you a real sense of accomplishment, at least within the game and among other gamers. Unfortunately, these accomplishments typically fail to translate to life outside the game.

Digging Up More Trouble Hidden within Video Gaming

Inside the video game world, many psychological factors account for the potency of the video gaming experience, as you find out in the following sections. This power may also contribute to the addictive potential of video games (covered earlier in this chapter).

Distinguishing social comparison from social connection

REMEMBER

How do you know when the connections made within game are simply based on competition that you experience with other players?

>> **Social comparison** is the evaluation of your performance through comparison with other players' performance in the game.

>> **Social connection** is intimate connection that is *not* based on comparing your performance to other gamers.

Most of the patients I've treated report feeling connected to people in the game, but only to a point; I'm not sure that this connection extends out beyond the game world. I can only judge based on what my patients share with me, but it appears the levels of intimacy are somewhat circumscribed around gaming and gaming-related activities. These relationships may exist *only* within the context of gaming, although some players report lasting and meaningful relationships that extend outside of gameplay. This is not to say that the video game addict does not feel a sense of belonging and connection to members of their gaming group. In my view, most of the social connection experienced with fellow gamers seems to be related to shared gaming activities, but some gamers report using apps such as Discord or their phones to communicate occasionally outside of gaming. (See the earlier section "Connecting on a shared challenge" for more about connection via video games.)

Putting in time to keep your edge

Like all competitive skills, video gaming requires maintaining your edge and has no easily definable end. Competition within a game can be fierce. There are many very skilled players — all of whom are motivated to keep advancing through the ranks of the game. To maintain your in-game rating or rank, you must play enough to both keep up your skills and demonstrate loyalty to the cause (the game). This can also foster addictive play because unless you want to lose your hard-earned place, you must keep playing.

To remain highly ranked or rated with a particular video game, you must put in your time and pay your dues; this is a rite of passage with all groups and organizations — to have credibility (and a sense of membership), you must earn it.

Seeing the power of role playing

When I did my original research in the late 1990s on Internet and video game addiction, one of the factors I looked at regarding Internet use was the attractiveness of role playing and the ability to take on alternate identities. There is a great deal of appeal in being able to relinquish your identity and take on an alternative identity, and the ability to experiment with using such alternate realities is very compelling. Some would agree that experimenting in this manner within a virtual world can be an expansive experience; in a sense, it is analogous to the *depersonalization* people experience with mood-altering and psychoactive drugs. With video gaming, this is done without ingesting anything — all you need is the time and willingness to take on new aspects of your personality and behavior.

Taking on new roles and personalities can be fun and exciting, and many games allow you to look and act differently, including how you dress, your gender, and the adoption of powerful abilities. It can be both intoxicating and thrilling to be seven feet tall, have horns, and be able to fly. This is fantasy enacted on a new level as you can see the character you create and feel like you *are* that character on the screen. It is probably the closest you can come to really being another person (or creature, as the case may be).

Interactivity also contributes to video gaming's potent addictive power. You are not only playing an addictive and stimulating game, but you are also playing the game with *other* like-minded people. The power of being able to play with anyone from anywhere in the world as if they are in the same room with you can be quite compelling.

Exploring e-sports

E-sports offer an new slant on the 50-year-old world of video gaming. Competitions are not new to video gaming, but the level of prizes and prestige and official competitions have grown significantly in the last ten or more years. E-sports have gone mainstream with pay-per-view and televised world competitions, sponsorships, and big money prizes. For gamers, this is much like watching the World Series or the Super Bowl — you can only dream of being at the top of the game and experiencing the fame and recognition of such high levels of achievement and potential money.

Is video gaming a sport? That is a very good question, and it depends on what you define as a sport. If your definition of a sport is a physically strenuous competitive activity, then video gaming probably does not qualify; however, if you define a sport as some form of competitive play with spectators, then video gaming may certainly qualify. The entire e-sports movement is at least partially driven by the fact that regional, national, and international competitions are good business, especially if they can be broadcast, streamed, or televised. There are big audiences and big sponsorship dollars for these competitions, and I cannot help but think that the e-sports movement has been at least partially advanced by the introduction of monetary rewards.

WARNING

To me, it matters less whether video gaming is defined as a sport or not; even if it is a sport (by whatever definition you use), it seems that the bigger issue at hand is whether the sport is *addictive* to the point where there are negative life consequences. I have never seen a person who hoped to become a professional player in a sport they loved devolve to the point of dysfunction in the way that I have seen with video gaming.

The one negative for defining video gaming as a sport is around the rationalization and justification for heavy levels of play. Because sports have a positive connotation, video gamers will use this as an excuse for their inordinate amount of play. In my years practicing addiction medicine, I have never seen anyone addicted to a traditional sport (although I have sometimes seen excessive play). This is not the case for video games, where I have seen many examples of addictive use.

Just like generations of kids who wanted to "be like Mike" by playing like Michael Jordan, gamers will emulate their gaming heroes and aspire to their level of achievement. How do you get to this level? You play, play, and then play some more. All this play can then be justified by the goal of reaching the pinnacle of video game competition. Unfortunately, just like in the NBA, very few kids ever make it to the playoffs — in fact, the odds of making it to that level are astronomical. Nevertheless, this belief in being able to get there can contribute to video game addiction and add to rationalization and justification of extensive play, and in some cases, I have had gaming addict patients use the goal of becoming a professional gamer to justify addictive play to their parents and families.

WARNING

The play required to emulate the famous pro gamers takes *enormous* amounts of time. All this time must come from somewhere, and typically it is stolen from sleep, self-care, academics, and social relationships.

Looking at loot boxes and paid treasures

Here is where video game addiction becomes problematic in terms of real-time costs. Many games have what are called in-app or in-game purchases. You can purchase (with real money) new and better weapons, level changes, new equipment, or an opportunity to spin a wheel for a chance to win something big. Every game is different in terms of what it offers, but all the offers imply a better opportunity to advance within the game. Most of the money that I have seen spent (and I have seen amounts as high as $10,000) has been on such in-game purchases. If you can pay money to get an edge up in the game and advance, why wouldn't you? In some cases, my patients have stolen credit cards or bank account numbers or used other indirect means to get money to pay for what they felt they needed in the game.

Loot boxes and in-game paid treasures are very appealing. They are colorful, loud, and attractive, and they move around in the game. This stimulation is very captivating and alluring to the senses, and this can further contribute to the addictive experience by promising a shortcut to game advancement and success.

Freemium games are a clever derivation of video games that are typically played on PCs, tablets, or smartphones. The game is essentially free to play, but only up to a point. To get to the next level, to get a new power or ability, or to simply play longer, you *must* pay a fee. Sometimes these are called *pay-to-play* games, and they are a sneaky way to get people to pay for what is initially free. Only about 10 percent of users end up paying for the extra time, access, or tokens, but that 10 percent is enough to make the games immensely profitable. I would say that there are undoubtedly some gaming addicts among this 10 percent. Candy Crush is a good example of such a game. The truth is that *nothing* is ever free; someone, somehow, is paying for it, and when it comes to freemium or pay-to-play games, a mixture of your time and money is required to pay the bill. This is an interesting phenomenon as the game may still *feel* free, even though you are paying for access to a loot box, treasure chest, or other in-game, play-to-play goodies.

Appreciating the concern over violent games

Violence in video games is a very controversial topic and the focus of significant research and debate in the last few decades. Although there is some debate within the behavioral health and psychiatric community regarding video game violence, numerous respected organizations such as the American Psychological Association have commissioned white papers and studies on the subject. Although there is no way to clearly tie violence in video games to later enactment of violence, there is little doubt that exposure to, and involvement with, video game violence has an impact on the minds and behavior of those who consume them.

Desensitization to violence

The current research in media-based violence, which certainly includes violent video games, has been quite extensive. First-person shooter (FPS) games are phenomenally popular and constitute a sizable percentage of some of the top video game titles. Grand Theft Auto, Call of Duty, and Halo all enable you to act as if you are holding a gun, sighting the weapon, hunting your prey (or victim), and then shooting. This process has become very sophisticated in terms of the graphic details of the violent activity. Most games have adjustments for levels of gore, from mild to extreme. In the most extreme modes, you can see the kill shot and the blood and gore in very realistic detail.

WARNING

So, what is the effect of all this violence? Well, the research seems to agree that although we cannot directly connect the propensity to commit violence with violent video game use, there does seem to be a clear connection with desensitization to violence from the violent video game experience. Some of the earliest FPS violent video games were developed by the U.S. Army to be used for both recruitment and training purposes. The specific purpose of the game was to help train recruits by desensitizing them to shooting violence, and they were, and are, very successful at doing just that.

The building of skills of violence

The other area of impact of video game violence is in skills acquisition and development. The game can teach you how to shoot, and how to master shooting better. While it may desensitize you to violence, it is also teaching you how to hold, sight, and shoot the gun accurately. The brain can learn tasks via this type of virtual rehearsal, and it is quite effective in its ability to transfer to the real use of a gun. There are numerous examples where people learned how to shoot a real gun accurately by using a video game. The same goes with the acquisition of other complex motor tasks, such as flying an airplane. It's not enough to learn the skill in a game simulation, but it gives you a significant head start and a significant amount of confidence in a complex set of behaviors, like shooting a gun.

WARNING

Another factor of violent video games is the impact around the glorification of violence, and many of these FPS games use sophisticated and interesting-looking automatic weapons. The idea that the portrayal of violence is entertaining appears to have some negative impacts on the psychology of heavy users of violent video games. Part of the basis for this effect is the fact that the brain is not great at separating actual behavior from fantasy; this does not mean that playing a violent game makes you violent, but it does mean that you are activating aggressive centers of your brain when you enact such game scenes.

Chapter **8**

The Risks of Online Shopping, Gambling, and Stock Trading

have always said that the power of the Internet is in its ability to efficiently connect people with other people, services, products, and information. When it comes to the Internet, *convenience,* not necessity, is the mother of invention. The Internet is the ultimate convenience tool. When it was integrated with the smartphone a mere 12–13 years ago, a new ability to always be online was born, and this advance has had huge implications for our health and well-being.

When it comes to anything connected to the Internet, ease of access is the main operating principle at work. All activities, such as shopping, dating, stock trading, banking, auction bidding, gambling, social media, and smartphone games, utilize the convenience factor in *easily* connecting people to the services and products they want. People like easy. In this chapter, I go over the risks of engaging in three activities with easy online access: shopping, gambling, and stock trading. Before that, I discuss an important aspect of these activities, cybersecurity.

Taking on Cybersecurity

What the heck is cybersecurity anyway? Many books can do this subject better justice than I can, but I cover the basics in this section.

REMEMBER

Cybersecurity is generally referred to as software, apps, and hardware solutions that attempt to limit nefarious exposure to bad things on the Internet. Malware, ransomware, viruses, phishing, scams, spyware, and hacks abound online, and there are many off-the-shelf software packages that will serve you well; some of these come with your device's operating system, while others you will have to buy as aftermarket products.

WARNING

Why am I talking about cybersecurity in this chapter? The reason is simple: Any transactions that include identity or money are very vulnerable to cybersecurity concerns. Programs, websites, and apps that deal with financial information and transactions often have encryptions and various hacking protections (although these aren't perfect either), but the biggest risk around cybersecurity when it comes to shopping, investing, and online gambling is the human factor (which I go into detail later in this chapter). Scams, fake websites, and deception thrive on the Internet, and it can be easy to fall prey to these breaches without even realizing it. When dealing with a purchase, an investment, or online gambling, sensitive financial information needs to be entered and exchanged — and this is a perfect opportunity for scammers to garner your personal data.

Yes, your computer can get sick with a virus (and more)

Perhaps the most common cybersecurity issues are viruses, and these are the most common computer and Internet bugs you can get. Some are automated, and some are more custom-designed for specific purposes. Some viruses are simply nuisances, while others can really ruin your day.

Malware is simply a group of bots (automatic robot-like programs) and software applications that either take information from your machine or put information into your machine. The most common use for malware is for marketing and business, to get information about you and your device that can be used either to sell you something or to sell your name and information to someone else. The word *malware* tells it like it is — it isn't looking out for your best interests. By the way, *ransomware* is a particularly nasty version, where hackers take control of your computer or your data and refuse you access unless you pay them money. This can really ruin your day and maybe your year.

In my opinion, it's impossible to be up on all the technical aspects of computers, smartphones, and the Internet, and having professional IT or tech support is money well spent. Some of the simplest solutions, however, are to have all the websites and apps that you use online be *password protected.* You probably know the joy of having to constantly change your passwords all the time, but good passwords are required to be more and more sophisticated as the threats increase. All of this does eat up your time and attention, but it's the negative side of the Internet and computer-based world that you now live in.

REMEMBER

It is far better to spend a few dollars for some good malware and anti-virus protection software and to know your computer and other devices are protected than to worry about something going wrong.

THE HUMAN ELEMENT OF CYBERSECURITY

Of course, you're always hearing about identity theft, online scams, cybersecurity, cybercrimes, and Internet safety. The truth is that the Internet's strength is also its greatest vulnerability. Because information is easily transmitted and received online, it makes hacking rather easy for those who know how to exploit the Internet's weaknesses. It's impossible to make computers and smartphones hack-proof. While malware protections, anti-virus software, and other measures are available to make being online as safe as possible, one thing cannot be made safe: *you,* the human end of the equation. Many, if not most, online scams have started with a deception that the Internet interface easily allowed.

I recall one such example that happened to me just recently, where I received an official-looking email notice from PayPal; it was on their letterhead and looked like the real deal. It said that my account had been hacked and that I should call the phone number provided in the email. This just so happened to coincide with my credit card being breached (a bad week for me), and I assumed the events were related because I thought my credit card was connected to my PayPal account. They were not related.

I called the number, and 15 or 20 seconds into the conversation, I was already beginning to type into my computer screen instructions from the kindly gentleman on the other end of the phone. He was simply trying to help me, after all. Twenty seconds into this process, my brain kicked back in, and I realized it was a scam. Had I completed his instructions, he would have had access to my entire computer's contents, and there was nothing I could have done about it because I would have invited him in. The entire event took less than a minute, and no software would have stopped this near-disaster, because the most vulnerable element in the equation was the human one: me.

The power of the Internet is always limited by the human factor. The Internet is amoral. It has no intelligence or conscience, and so whatever conscience it may have comes from *you.*

There are many good anti-virus and malware programs and apps. Most people have some sort of anti-virus software on their devices; make sure that you have this software and that it is up to date. Also, it's always a good idea to have a computer tech who is knowledgeable about cybersecurity to make sure you're protected. Ask your tech person what they like, use, and are familiar with. Sometimes it's best to use what's familiar to whoever is going to be working on your devices as opposed to one that's "rated best" by a magazine or other source.

TIP

Where do you find a computer or smartphone tech? Most stores that sell equipment or services also offer tech support and IT help. Best Buy, Staples, and most of the mobile phone service providers have some tech support options available. You can also find many freestanding shops that specialize in computer and screen device repair, and often they are quite knowledgeable and have IT experience. I'd say look in the Yellow Pages for IT or computer tech support, but there are no more Yellow Pages, so yes, do a search online for local services and IT providers that can help you. Again, it's impossible to be an expert at everything, and sometimes you need help in making sure you're reasonably protected.

Many people who have some tech knowledge fancy themselves tech experts, and they may in fact know much more than most people, so they can be great resources if they have the time, but sometimes, especially if they are family members, that can create other complications. Other times people may not know as much as they think, so getting professional help can really be worth the money. Spending a few dollars can often save you a lot of headaches, and potentially money, in the long run.

Don't trust, and still verify

As I say earlier, perhaps the most vulnerable aspect of the Internet equation is the human one. Most of my worst disasters or near-disasters — like the one I mention in the nearby sidebar "The human element of cybersecurity" — were from my carelessness or simply my *humanness* in trusting another person. Viruses, malware, ransomware, and scams are plentiful online, and it would be good to remember the adage "If it seems too good to be true, it probably isn't." This goes for online shopping as well. Case in point: I bought a set of EarPods for my son recently for what appeared to be a good price on a website — the problem was they were fake. They looked real but they were not. I could give you more stories about how what appeared true was in fact not. This is the Internet's biggest liability: the veil of quasi-anonymity that cloaks the truth of what is being presented.

WARNING

Trust is earned, and online you really have no way of knowing whom you are talking to or dealing with. Take care not to give out personal information to anyone online. No legitimate person or organization will ask for such information. Online anonymity sometimes shades the identity of whom you're dealing with, and even with legitimate-looking logos and emails, it's possible that you're talking to an imposter who is simply trying to get your information or money.

REMEMBER

As they say in government, *trust but verify,* except in this case, I am saying *do not trust and still verify.* Even when I am speaking, emailing, texting, or chatting online, I am always expecting the possibility of a scam. Take care and do not be afraid to ask how you know who you are talking to online; ask for verification and proof. Someone legitimate will not mind and will gladly comply. It is far better to assume a scam than to be taken by one.

Shopping Online: The Socially Acceptable Addiction

Online shopping, the subject of this section, is a topic close to my own heart in that I shop almost exclusively on Amazon. (This is not a plug for them — God knows they don't need it.) The percent of total U.S. online retail sales for Amazon for 2020 was a whopping 47 percent! Total online retail sales for Amazon in 2021 are likely to be 15 percent of *all* retail sales in the United States. Amazon is by far the leader, with nearly 50 cents of every U.S. online dollar going to Jeff Bezos's company. This is quite shocking in that Amazon is barely 25 years old, having started in 1995.

The point of all this is that in 25 short years, the entire retail sector of the United States, and most of the world, has shifted. Why? I give an answer to this question earlier in this chapter: When it comes to the Internet, *convenience,* not necessity, is the mother of invention. My case is a good example. I have some physical limitations, and it is far easier for me to order an item on Amazon that I do not immediately need. I wait a day or two and then open a box, as opposed to getting in my car, finding the store, hoping to find a parking spot, hoping the store has the item, buying it, bringing it home, and hoping I do not need to return it and start the process all over again. I've never added up all this time, but I imagine it is quite a bit. I order from Amazon and other online retailers because it is easier and allows me more time for other things.

But there is a dark hue to this e-commerce rainbow. Online shopping can easily become addictive. It is so easy to *pick* and *click* on Amazon or another retailer that you forget that all this perfect consumer design is aimed at the limbic reward system in your brain (see Chapter 2 for more about your brain's biology). It is like having Christmas 365 days a year, and the fun of getting endless packages all the time, with all sorts of goodies, can clearly become intoxicating and addictive. I have not seen the data on this yet, but I am sure that I (and likely many others) spend more money on Amazon than if I waited to buy what I absolutely needed in a brick-and-mortar store. Somehow the allure of that endless choice, instant gratification, and the two-level dopamine hit from buying and then opening the item acts like a consumer drug.

REMEMBER The easier it is to get something that elevates your dopamine levels, the more likely it is that you will keep doing it.

WARNING In my years of treating Internet addictions, I have rarely seen online shopping as an initial presenting problem, but often as a problem associated with video gaming (which I cover in Chapter 7). In video gaming, the gaming addict can get very caught up in buying in-game items to a point where they might be spending thousands of dollars. I have seen a few serious cases of online shopping addiction, usually to a point where the addict had to steal from family to feed their addiction. I suspect that many of us overspend online, but we are only just beginning to look at this data now.

Part of the hidden aspect of online shopping addiction is undoubtedly the socially acceptable aspect of shopping and consumerism; you take it for granted that shopping is fun, but you don't see the more extreme examples of this behavior, except for the hard-core hoarding behavior you might see on TV. Online shopping easily supports this type of hoarding addiction because the packages come to your home with little human contact, and it doesn't take long to fill a house with these addictive cyber-spoils.

Trading money for convenience

REMEMBER It is not always true that online shopping saves you money (see the previous section) because in my case, I am certain that I buy more things than I would otherwise because of the ease of access, availability, and near-instant gratification. This is where you must take care to avoid the addictive elements inherent in online shopping, including spending excessive time scrolling and browsing online. Try to set budgets of what you can reasonably spend, consider whether you really need the item or not, take care not to shop online when you are tired or in a negative mood, and lastly be cautious of what I call *consumer bulimia:* the tendency to shop and then immediately return things out of guilt or realizing you spent too much money. Online retailers take advantage of your eyes onscreen while they have you there, and they will tailor what is on your screen to fit what they *know* you like or what you have looked at before. They know that you are *more likely* to buy an item after you have been exposed to it numerous times.

An auction site presents an even more potentially addictive experience because you may or may not win the item you're bidding on. What does that sound like? It is exactly the way the *maybe* factor works with a slot machine. I recall spending more money than I ought to have on eBay in the early days because of this factor, and now I see the same potential, perhaps to a lesser extent, with Amazon. Amazon often tells you how many items they have on hand, implying that you should order now before they are gone; you have no idea if those numbers are real or an attempt to instill urgency and prod you along to buy now. Either way, it works.

We know that ease of access is a significant factor in the addictive process. Shopping is also an easy addiction to hide as you can rationalize that you need all the items you're purchasing. However, this isn't the case, and I know that Amazon and other online retailers count on the fact that you'll buy things on impulse at checkout just like you do at the endcaps of the grocery store. (By the way, the sale of impulse purchases at the endcaps of a store often accounts for a sizable percentage of profit for the store.)

Getting a hit of dopamine

I know that shopping has many of the same mood-altering properties that other drugs and behaviors have in terms of dopamine. Although I am not comparing shopping to using cocaine or even video gaming, you know the excitement of finding something you want, purchasing it, opening it up, and trying it out. This dopamine hit is generally short-lived and often leads to more purchasing and shopping to continue to find additional hits. Hence the consumer cycle. There is no question that this can become an addictive process, and I have seen this in my own life as well as with my patients, whether on Amazon, eBay, or other online shopping sites.

Any drug or behavior that is *short-lived* in its dopamine hit is more likely to be repeated sooner, and the shorter the time from the click of the mouse to the arrival of that package, the more addictive it can become. So, as Amazon and other retailers speed up their delivery systems, this will undoubtedly increase the potential for addiction to shopping online.

Everybody loves to shop — well, maybe not everybody. But everybody loves to get that new thing they have been wanting. You may not like the process of going from store to store, but online, you can do all your shopping in a fraction of the time. The act of looking for and finding what you want is very dopamine elevating. The anticipation of finding, buying, and waiting for the package is a hit unto itself, and when it arrives, you experience a secondary hit of dopamine. Shopping and gambling (covered in the next section) indeed have much in common.

I mention in the previous section that auction sites are particularly addictive, and I give some examples of how you can get caught up in the online allure of winning a prize. And that's exactly what it is — winning the prize. There is a competitive aspect to winning the item of your dreams in an online auction. First, and perhaps foremost, the more people who want what you are bidding on, the more *perceived value* that item has. The auction house or seller wants that competitive factor because it's likely to drive the price up, and in many cases, you can pay more for an item than what it's worth. *Maybe* is always addictive, and auction sites amp up the *maybe* factor.

Placing a Risky Bet with Online Gambling

Gambling addiction, the topic I discuss in this section, is an identified medical disorder that has been known as an addiction for decades. Currently, it is labeled as *pathological gambling,* and the act of gambling in any form can be quite addictive. Even if you are not a gambling addict, you can feel the addictive pull if you walk into a casino, venture to a racetrack, or watch the lines of customers hoping for a big jackpot on lotto day. Gambling is just plain addictive because *it uses variable rewards* (for more on that, read Part 1). All gambling does this, but the Internet allows you to have instant and easy access to gambling sites at the click of a button. Winning is always unpredictable but *possible.* It's the possibility of winning that drives the dopamine reward circuit. For all types of gambling, you typically do win in some form on a regular but unpredictable basis, but on average, odds prove you will lose in the long run.

With gambling, the *maybe* factor is so strong that I've seen adults at the craps table do all sorts of crazy things to the dice before they throw them; they insist that these behaviors will make a difference on the outcome of their roll. This is superstitious conditioning at best, where at one point their number happened to come it when they blew on the dice in a certain way, so they now assume it will help again. Humans are highly susceptible to this phenomenon, and even though we logically know the dice have no memory, we believe that if we roll a certain number, that could affect the next roll.

WARNING

It should be noted that online casino gambling is currently illegal in the United States, but there are ways to access offshore gambling sites, and people run up huge debts through offshore gambling casinos. These sites are extremely dangerous, and you can use credit or debit cards or direct bank transfers to fund your gambling. I would never recommend using these sites as they are offshore businesses, and the servers are not housed in the United States; if there is an issue, I think you will have little recourse. The combination of the dopamine power of gambling, along with the ease of access and anonymity of the Internet, makes these sites too tempting to be safe to use.

A perfect (and dangerous) storm: The Internet and gambling

Throughout this book, I talk about how the Internet is the world's largest slot machine, and with gambling, I am adding another addictive behavior to an already addictive medium. The result is a near-perfect means by which you can anonymously place bets (in some cases illegally) with no threshold to cross, no one to examine what you're doing, and no one to question you. Admittedly, this low level of scrutiny is what makes online gambling so attractive and so dangerous.

Many offshore casinos use credit cards to allow you to bet with, and certainly on the dark web, you can gain access to all sorts of virtual betting with all the bells and whistles of the real thing. I have seen patients run up huge debts very quickly. Whether they're on the web or the dark web, these sites can be way too tempting for an Internet gambling addict to resist because of the ease of access and anonymity. The offshore casino sites look and feel like the real thing, and they lure you with the colors, sounds, and feel of being in a real casino. These sites and sounds create a *dopaminergic trigger experience*, much like a notification on your smartphone; they let you know there is something potentially good waiting for you.

REMEMBER

The power of the Internet is that it allows you to connect with the services, information, and products you want. It does this easily, efficiently, and somewhat privately. Online gambling is a perfect way to gamble because it allows access with little or no interference, and all with your credit or debit card number.

Offshore casinos do not have to conform to U.S. gaming regulations, and online bookies certainly do not either. There is a huge underground economy, and with the dark web, credit cards, wire transfers, Bitcoin, and other cryptocurrencies, you can get in over your head rather quickly. In the privacy of your own home, you can play as if you are at the casino. Of course, you have no way of knowing if the casino and games are fairly run, but even if they are, statistics show that if you play enough, you will eventually lose.

WARNING

The dark web (see Chapter 1) is the Wild West of the Internet where you can find anything and everything, legal and illegal. Here you can gamble in any number of ways, and you really do not know whom you are dealing with or where they are. I advise staying away from anything on the dark web.

Playing just to play, not to win

Legal lotteries can be played online; for example, many states have online and computerized Keno games that are computer generated and run all day and night. All you have to do is sit there, pick out a number, and hope you win. The fact is that you'll win sometimes; occasionally, you'll pull out some winning numbers and win back all or part of your bet. So, you bet again; after all, you're now using the house's money. The problem is that you're still gambling, and ultimately, the small intermittent wins are enough to whet your appetite and keep you gambling some more.

Gambling system designers understand the human brain's reward system very well, and all they need to do to captivate the primitive reward system in your head is to provide some wins, some small hits of dopamine, and the fuse is ignited. This kindling effect in gambling is only amplified within an online interface.

REMEMBER

Most people assume that you play a gambling game to win. This is only partially correct. The more accurate statement is that *you play so you can keep playing.* You see, most of the fun is in the playing. Most of the neurochemical reward is in the anticipation of possibly winning. Sure, winning feels great. But if winning were the end-all-be-all, then when you won enough, you would stop. That is not typically what happens. Winning is good, but winning *more* is better. The basic rule of all addictions is that if one is good, two must be better.

Professional gamblers typically stop playing *both* when they win and when they lose. They have a systematic method that, on average, allows them to make a living. They are not chasing a hit. Some professional gamblers (and perhaps professional video gamers as well) start out as addicts but learn that in order to function, they have to manage their addiction.

REMEMBER

Chasing the "high" phenomenon is exactly what we see in all addictions. Even with the Internet, you're always looking for something that will give you that hit of excitement (and dopamine), and the anticipation of finding just that perfect experience compels you to keep playing, surfing, or scrolling for excessive amounts of time. The Internet truly is the *world's largest slot machine.*

If you ever find yourself spending a huge amount of time looking for that perfect video, image, posting, photo, story, thread, or whatever, you are gambling. Your brain is looking for something, you may not know what, but it knows that somewhere lurking in the trillions of bits and bytes on that screen is something you will like. All you need to do is look (play) long enough and you will find it. That is the power of the Internet — so much so that in my original research in 1998, I adapted the diagnostic criteria for *pathological gambling* to study the then-new phenomenon of Internet addiction.

Rolling the Dice with Online Investing

In this section, I talk about stock trading and online investing. Now, how can that be a problem? Investing money in the world of finance is a science that is only enhanced by its use of the Internet, isn't it? Well, most honest investors and financial professionals would have to admit, if hard pressed, that all investing is, in essence, a form of educated gambling. The operative word is *gambling,* however, and the same rules for gambling operate with online stock and investment trading.

Just like when the porn industry (covered in Chapter 9) discovered the Internet, there was a revolution when the investment industry capitalized on the Internet to attract new customers and to service old ones. The ease of access online allows trades to be made instantly and efficiently and, I might add, with *little thought.* One of the problems with online investing is that you can pick and click a stock or other investment and, in five seconds, own something that may or may not be in your best interest. This is not to say the financial industry is trying to make this happen, but they, like all the other online businesses, want to capture more of the market and give their customers what they want and need.

Clicking before you think — over and over again

Consider the example of the professional gambler who uses gambling as a tool to make a living but is not swayed by the *pain of losing* or the *joys of winning.* The same needs to be true for the online investor. You must recognize that the Internet medium is addictive. I have seen many people thrown off by the ease of clicking on something without really understanding that the purchase they just made was essentially a form of gambling. Stocks, bonds, and other investments are not products, even though many financial professionals call them that. They are opportunities, and as opportunities, they may or may not bear fruit.

Please don't get me wrong; I use online investing and online banking. I am as wired as the next person. I like all these online options and services. But I clearly see the dangerous lure they provide. They speak about convenience, but sometimes that convenience means you click before you think, and if it turns out well (and sometimes it does), then you think the next time you click will give you the same results. And because online trading is done directly with your computer or smartphone on the Internet, your trade is devoid of human contact. In many cases, no one sees your trade, and certainly no one is going to call you up and say, "Mr. or Ms. Jones, are you sure you want to make that trade? It is awfully risky."

WARNING

You can, and do, lose even if you invest wisely. Online trading is as risky as offline trading, and perhaps more so, because you have that instantaneous hit of excitement at the anticipation of buying the next big winning stock. Online and offline investing, if done smartly (I am not offering financial advice here), is about mitigating risk — which is a fancy way of saying balancing risk, so that sometimes you win and sometimes you lose, and hopefully at the end of the day (or years), you come out ahead. But the pick-and-click ease of the Internet may convey an entirely different idea that all you have to do is pick a good stock and you are all set. I remember that when I was first introduced to online trading, I thought it was great, but at the end of the day, I made a lot of mistakes in my near-addictive enthusiasm. I now have professional investing help.

Equating investing with gambling

If you ask for an honest opinion from an experienced investor whether online trading is like gambling, they will most likely say it is. Keep in mind that the power of the Internet is in its ability to allow to you to do things quickly and easily; sometimes that does not allow for a lot of thought, and emotional choices are, in a sense, a form of gambling. The Internet can be used as a tool to help place educated and reasonable bets on stocks and other investments. But notice that I said "bets." Nothing that you invest in online is a sure thing. The best you can hope for is an educated guess and the wisdom to know the difference between a *guess* and a *sure thing.*

So, can you really be an online investing addict? The answer is yes. They typically do not present for help, but I have seen and treated people who had an issue with their use of online investing sites.

REMEMBER

The real power of the Internet is in its convenience and its ability to connect you with what you want or need. Don't let it fool you. It is, in and of itself, dumb. The Internet knows nothing. It only knows how to execute your uploaded requests for whatever you want and to download something back for you. There is no real intelligence behind the Internet in its current form (there is intelligence in the people who construct the content and algorithms). When I speak to Alexa and ask her what time it is (notice I said *her*), the reality is that she is faking an intelligent answer. We imbue the Internet with sentience. Be careful not to do so too much, lest you forget that there is no human behind that curtain.

Chapter **9**

Combining Addictions: The Power of Porn and the Internet

The history of the relationship between pornography and technology is long and entwined; from the earliest days when humans depicted sexual images on cave walls, to the modern use of the Internet, technology and sex have always co-existed. The use of technology to record, express, and distribute sexual depictions and behavior seems to be part of human sexuality, and some of these technologies have helped facilitate addictive and compulsive use of pornography.

The prevalence and current availability of online pornography is startling. According to the Shift Project, recent 2019 estimates state that approximately 27 percent of all videos played online are for porn; this is a staggeringly large number when you think about the total amount of video content consumed online. Recent surveys suggest that 33 percent of all Internet searches (across all search engines) are for porn. In 2018, carbon emissions generated from video servers and screens viewing porn were equivalent to the entire residential sector of France, and this represents 4 percent of all carbon emissions associated with Internet technologies across the globe.

In 2017, Pornhub (the world's largest porn website, which owns numerous other porn sites as well) described having 81 million visits per day, and that number is likely higher today. In fact, during the 2020 COVID pandemic, online porn use spiked even higher. In 2017–2018, Pornhub (only one of many online porn sites) reported 25 billion porn site searches. Although most estimates of Internet pornography users are predominantly male (90 percent), Pornhub reports users to be 25 percent female, although this is all self-reported data, so it's uncertain how accurate those numbers are.

Porn is also big business, with recent estimates as high as $20 billion, and many of the companies that produce and distribute porn are traded on public stock exchanges.

A large percentage of Internet porn is view by adolescents and young adults, with estimates that 90 percent of children ages 8–18 have seen online pornography, and with only about 60 percent of Americans receiving formal sex education, some argue that online pornography has become a major source of sex education for our youth.

REMEMBER

This chapter offers a medical and scientific discussion of a sensitive topic that has social, moral, ethical, religious, and values-based implications. The purpose of this discussion is to offer information and educational content — social, moral, ethical, and religious values will not be addressed. For the purposes of this chapter, I am offering no judgment on such values-based considerations, but they are outside the scope of this book. This chapter includes an explanation and discussion of sexual behavior and pornographic content, and I approach this material from a scientific view of human evolution, biology, and sexuality.

Understanding Why Pornography Is So Appealing

To understand why pornography is so appealing, it's important to understand the history of porn and how it has frequently been integrated with technology. Some depictions of pornography date back over 25,000 years, and there are clear examples of graphic drawings, paintings, prints, and carvings that show early detailed depictions of human sexual activities. There are even examples of hieroglyphic and petroglyphic drawings and carvings on cave walls that clearly show men and women engaging in sexual acts. Why? We seem to have a long history of fascination (and probable arousal) in the recording and expression of human sexual behavior. One might argue that the desire to engage in such depictions may be directly related to human sexual drives.

REMEMBER

If you examine the reward structure and circuitry in the brain (see Chapter 2), you clearly see that the dopamine reward system evolved for a reason. As with most aspects of human biological evolution, that reason has to do with survival. There are perhaps no greater examples of survival behavior than human sexual activity and procreation. Sex and survival are inexorably linked from a biological perspective. From the brain's perspective (if you will), all sexual behavior is linked to procreation and the importance of extending our genes into the next generation. That is not to say that everyone needs to have children; today, we have many choices about our sexual behavior, but from an evolutionary time perspective, our brains have changed little over the last 25,000 years. To the brain, all sexual behavior is linked to a survival drive (although other parts of the brain can and do mediate those drives).

So, what does that have to do with pornography? The likely reason we evolved a pleasure response connected to dopamine innervation seems to be to ensure that we engage in survival-related behaviors. Sex and food are two of the most potent and necessary aspects of our species' survival. It's my contention that pornography, and the tendency toward interest in sexual imagery and content, are directly linked to this survival drive. That is not to say that we must view pornography, because certainly we do not have to. Rather, I am saying that the desire to enjoy and want to see such images is likely linked to the primitive reward structures in the human brain. What this demonstrates is how we can become addicted to viewing such images. Like all addictions, porn addiction is piggybacking on those original dopamine reward features of the limbic system in the brain.

In the following sections, I cover the basics of pornography's addictiveness.

The potency of pornography even before the Internet

In ancient renderings of pornography, the spiritual and religious were often integrated into erotic imagery. During Greek dominance of the Western world, this spiritual influence disappeared, and the imagery seemed to move toward more hedonistic and graphic depictions. Modern pornography might similarly be described as erotic imagery intended for sexual arousal, devoid of social or spiritual components. It is likely that this more hedonistic use of pornography may contribute to some of the drug-like, addictive properties related to the consumption of Internet pornography. If you remove the *nutritive* elements of a pleasurable behavior, you may pave the way for more compulsive and addictive use.

In modern times, there is a split between sexuality and the sacred that can leave sexuality as shame-bound, and porn seems to reflect this covert separation. With pornography, you can experience potent sexual arousal devoid of intimacy or relationship — sex made simple.

Pornography addiction issues are the second-most common reason why people seek help at our clinic and are only surpassed by video game addiction (see Chapter 7) as a reason for seeking professional help. It is interesting to note that both video game use and porn use are addictive behaviors, even without the Internet. Both were shown to have addictive properties long before being linked to online platforms. What this says is that both behaviors have psychoactive qualities that elevate dopamine, and that when combined with an Internet delivery modality, it is like pouring gasoline on a fire.

Of all available online content, pornography may be the most stimulating and potent, as it neatly connects to your brain's reward system. Viewing sexual imagery in pornography certainly elevates dopamine, but it also kindles anticipatory dopamine release triggered by the hippocampal memory circuit, as well as other hormonal activity related to sexual arousal. Although this is difficult to quantify, the brain also associates that a further reward for viewing pornography *might* be followed by a potent explosive expression of dopamine from orgasm. Researchers compare the high dopamine levels in the nucleus accumbens, released at orgasm, to the high experienced from drugs, such as methamphetamine and cocaine. Here we have a two-tiered reward, one linked to pornography (arousal) and the second relating to orgasm — although not all porn use is followed by an orgasm.

I emphasized the word *might*, which is another way of saying *maybe*, and you know how addictive the word *maybe* is to the brain if you've read Chapter 3.

A BRIEF HISTORY OF THE RELATIONSHIP BETWEEN PORNOGRAPHY AND TECHNOLOGY

If you were to record human existence, you would be including sexual and pornographic content. It began with the ancient artistic depictions on cave walls and moved into art and sculpture, and there are numerous examples of such depictions in ancient Greek, Roman, Indian, and Far Eastern cultures.

After the Bible (1200 BC approximately), some early uses of the printing press (1440) were to reproduce sexually explicit writings (perhaps not explicit by today's standards, but racy for the time). This was later followed by another major technological development: motion pictures. Some of the titles of the earliest silent films had sexual content — certainly not pornographic by today's standards, but by the standards of the day, sexually explicit. For perhaps 100 years, pornographic films were crude and poorly produced, all the way up through the advent of 8mm and 16mm home movies. Although these movies could be viewed privately at home, most pornography was viewed in "dirty movie theaters," and when I say dirty, I mean they were filthy. From the 1950s on, these films could also be viewed in private *peep shows* that were usually in the back of adult bookstores (also a less-than-homey experience). But all of this changed with the invention of the video cassette recorder (VCR) in 1971. Both Sony and Panasonic pioneered the technology, and even early on, their videotape technologies attracted pornography producers who wanted distribute content on their new video cassette platforms.

This was a game changer because now people could buy or rent pornographic titles for private viewing in their own homes. Anonymity is a big driver of sexual behavior, and now they had it with the VCR. One story that illustrates the power of pornography's influence on technology is the VHS and Betamax story. You may recall (if you are over 40) that there used to be two main formats for videocassette recording: VHS and Betamax (also known as Beta). Beta was technically superior and was rapidly adopted by the broadcast industry, but it had a limitation of only one hour of content permitted on a tape. Although there were adult titles in both formats, it quickly became apparent that the porn industry preferred the poorer-quality VHS format; they could simply fit much more content on a tape. Some argue that this is one reason why the Beta format slowly disappeared from video store shelves.

The integration of pornography into the tech revolution did not stop there. The next major evolution was the appearance of video CDs (1982) and DVDs (1995), which the porn industry readily adopted; these mediums were soon followed by the porn industry's very early (and enthusiastic) adoption of the Internet (which is their current medium of choice).

The increase of online pornography addiction

The Internet (with its popular use ramping up in the early 1990s) allowed the best of everything for the consumption and viewing of pornography: privacy, disinhibition, interactivity, novelty, anonymity, endless choice, and ease of access. All these features helped create a perfect storm for the addictive use of pornography. As soon as access to porn became simple and easy, the number of addicted users also increased. Although no precise numbers for pornography addiction are available, to put it into perspective (as mentioned earlier), Internet use data shows that approximately 35 percent of all Internet searches are for porn or adult material. I can informally track this increase through my own 35 years of clinical practice — as Internet technologies improved, the presentation of pornography addiction and related cases in my practice has also steadily increased. No doubt, this increase contributes to online pornography and Internet sexual behavior being the second-most frequent reason for seeking behavioral addiction treatment. To further pour fuel on the fire, enter the smartphone (approximately 2007–2008), and we now have untethered and portable porn to carry in our pocket.

Typically, online pornography addiction follows a similar course in most of the patients I see. Although statistics show that women consume online pornography at a higher level than ever, the porn-addicted patients I see are exclusively male. The age varies greatly, from early teenage years (age 12 was the youngest I have seen) through their late 70s and even 80s. It seems that the desire to view pornographic imagery does not always fade as one ages and appears to be independent from testosterone level, circulatory functioning, marital status, or even real-time sexual activity. Indeed, I hypothesize that porn may stimulate a more primitive visual process linked to arousal and reward circuits embedded deeply in the brain.

Porn has morphed over recent years, owing to the Internet's advancing technology, and access to pornography now includes new hybrids of social media, influencing, and sex. Recently, there has been a further expansion of ways to produce and access porn through new channels on social media such as TikTok; here, content creators post revealing short videos and photos with links to fee-based, members-only access in apps such as OnlyFans in order to view complete nudity and hard-core porn. These new hybrids put content, production, distribution, and access all under the sex workers' or models' control. I have dubbed this new area of content *social-sexual*, as it combines the power and reach of social media with sexual content — with the most revealing content provided through a paywall, along with tipping. Webcam sites also invite private viewing, and some have sex toys (which are worn) that are linked to the Internet and are activated by people tipping money online. The more people tip, the longer the vibrations of the sex toy.

Addiction to online pornography has a varied presentation, but it almost always involves frequent and compulsive viewing of porn, often followed by masturbation (although masturbation is not always involved). The key is that the addict spends large amounts of time searching for, recording, and viewing pornographic images and videos. Sometimes erectile dysfunction (ED) can be a result of excessive porn use, and other times porn may become a preference over real-time sexual relations, even when in a marriage or other relationship. Typically, many people do not view their pornography use as a problem, at least initially, but after getting caught by a parent or loved one, fired, or in legal trouble, they might begin to recognize they do have a problem.

WARNING

Many of the notable issues and consequences that arise from porn are marital and relationship problems and getting caught at work (I have many patients who were fired from their jobs); others have major problems at school, including expulsion. In short, the consequences are often significant, with far-reaching problems arising in their lives due to their online pornography use. This is the hallmark of an addictive process — *the continuation of a behavior despite severe negative consequences.* Porn addiction, as with most addictions, is not logical. The only logic is the potency of the digital drug that is being overused and abused.

REMEMBER

All digital content (think of it as a form of digital drug) has the power to release dopamine; some content, such as pornography, is a very powerful releaser of dopamine, and therefore has very potent addictive properties.

The development of other serious problems from porn addiction

WARNING

In some cases, the porn addict will desensitize and develop a tolerance to the images and videos they are viewing and downloading. When this happens, they will sometimes try to find more novel images and videos to look at. This is where things can go terribly wrong. I have worked with numerous patients over the last 25 years (as an expert witness), who, once addicted to pornography, no longer received the same dopamine rush. These patients will, at times, gravitate to illegal content involving underage children and teens, which provides some of the arousal they were no longer getting from legal adult pornography. These cases often have severe and tragic endings with criminal charges and jail time.

It's important to note that in my experience, most porn addicts are not pedophiles (although most pedophiles use child pornography), and they do not typically have a sexual arousal template that is geared to children. However, what they do have is a porn addiction, and as with all addictions, a tolerance can develop that makes the previously dopamine-elevating pornographic content no longer satisfying, thus leading them to search for novel images.

Unlike adult pornography involving models and actors over 18, images that involve children and teens under 18 are *illegal.* Some unfortunate addicts end up in very significant legal trouble, which amplifies the serious impact of their addiction.

Identifying the Many Manifestations of Online Sexuality

There is perhaps no broader use of Internet technology than online sexuality. When the Internet was just heating up in the 1990s, there was no more explored use for this new technology than sexuality and sexual behavior. In 2000, I (along with several colleagues) published an article titled "Sexuality and the Internet: The next sexual revolution," which appeared in the journal *Psychological Perspectives on Human Sexuality.* The article explored the unique aspects of Internet-enabled sexual behavior. Even then, it was obvious that the Internet would change human sexuality in significant ways. And, like all changes, there have been positives and negatives to the marriage of sexuality and the Internet.

In the article, I discussed how perfectly suited the Internet was to both healthy and potentially unhealthy aspects of human sexual behavior. There is little question in my mind that the integration of the Internet and the adult sex industry (particularly the porn industry) is largely responsible for the increase in sex and porn addiction. The reason for this is simple. If you increase ease of access and make the activity private, you're going to increase use, and this will ultimately lead to greater amounts of addiction; I am sure that the porn industry does not acknowledge its responsibility for the problems associated with porn addiction, any more than the video game industry is doing so for video game addicts.

The ease of access, perceived anonymity, timelessness, and disinhibition create a near-ideal medium to engage in sexual activities of all types; this is not limited to pornography and adult video content, but also includes new areas of digital sexual behavior that have not been previously seen. They include sexting, webcam sites, live videos with sex toy interaction, self-directed pornography via the Internet, fetish sites, sexual singles and meet-up sites, social media–linked porn access, and even online directories for escorts and prostitution. It seems clear that the sex industry was transformed and expanded with the Internet.

I've had numerous patients who have come in after getting caught up in online sex scams, where a sex worker took advantage of the emotional (as well as sexual) distress of the person, eventually taking tens of thousands of dollars over extended periods of time. Although it was clear that they were being scammed, their addiction, along with its cognitive distortion and rationalization, prevented them from seeing the truth of the situation and recognizing the scam. The consequences of these situations are significant in terms of family disruption as well as the serious financial losses.

Sexting

What exactly is sexting? You may have heard this term used from time to time, and certainly it has been in the news over the last 15 years. Essentially, it involves sending sexual content (including photos and videos) via the Internet or a cellphone connection. Often this is done using smartphones, usually over Wi-Fi, but it can also be sent over standard 4G and 5G cellphone service — perhaps at a slower speed. Sexting may contain graphic sexual language, along with images and videos; it often contains personal photos and videos taken of the user and sent to another person and may or may not be reciprocated. Sometimes these photos or videos are uploaded to social media and other platforms, without the consent of the person in the images. Other times, these images can end up being distributed to other unintended people.

Sexting has a reputation of being used by adolescents and young adults. There have also been cases where children have sent sexual content over their phones or the Internet. But perhaps most surprising are the number of adults of all ages who have begun to send sext messages. Although there are fewer legal issues involved with adults, there can be some embarrassing moments if the photos are seen by the wrong person. It should also be noted that research suggests that not all sexting is problematic, and that among consenting adults, with good boundaries, it seems to be fine.

REMEMBER

Whenever you send anything via the Internet, including with a smartphone, you can *never* be completely sure whether the intended recipient is the only one who will see that text, photo, or video — *ever*.

WARNING

Legal issues are potentially significant when it comes to sexting. I am not an attorney, but I have dealt with enough of these cases to know that there is potential liability when a child under 18 sends a sext message or when a person over 18 sends a sext message to someone under 18. I have seen all these permutations, and there can be significant problems in terms of a person possessing sexual images of a minor and then sending them to someone else; this might be interpreted in the same manner that child pornography laws are interpreted. It is important to recognize that there are laws involved in possessing sexual or nude photos of a minor. Even if the photos are sent from one minor to another, as often occurs, there can be potential legal issues, although the courts have become somewhat more understanding in cases of two lust-filled teens with bad judgment. Nevertheless, care needs to be taken when dealing with sexual images sent from one device to another.

Dating and hook-up sites

It is estimated that one-third to one-half of all intimate relationships begin on the Internet. The sites and apps that users begin with are called *dating sites,* and there are dozens of them. Many are owned by a few key companies, and sites like Match, Tinder, Hinge, Bumble, Plenty of Fish, and eharmony are well-known brands, but there are many niche sites that cater to certain age, ethnic, or special interest groups. When it comes to dating sites, they have truly gone mainstream, and it's common to find many people who are dating, in relationships, or married who began their romantic life online.

REMEMBER

The reason dating sites are so popular is simple. Going back to the phrase I mention many times in this book, the power of the Internet comes from its easily accessible way of connecting people with the things (and in this case, people) they want. That is it. The most basic aspect of the Internet is what makes dating sites work so well.

One addictive feature that we see with dating sites is that people are in a way commoditized, and because there appears to be endless choices, the dating sites become almost like slot machines; some users become addicted to endlessly checking and searching on these sites, never really making a choice.

Beyond dating sites are hook-up sites and sex worker sites, which are covered in the following sections.

Hooking up while hooked up online

Hook-up sites are kind of what they sound like. They are designed to connect people who want to find willing sexual partners, well, for sex. There are many different types of hook-up sites. Perhaps the most famous is Tinder, which probably functions more like a dating site now that it has been mainstreamed. Grindr, for gay men, was perhaps one of the earlier sex hook-up sites.

Another perhaps infamous site is Ashley Madison, which essentially advertises a guarantee to help you find someone to have an affair with. There are others of the same ilk, including Fling and Adult FriendFinder. They all purport to do the same thing: help those who want sex to find sex. These sites typically are not offering professional sex worker services (although some of the ads are placed by professionals looking for customers).

One interesting side note about Ashley Madison has to do with cybersecurity (which I discuss in Chapter 8). There was a point a few years ago that their servers were breached by a hacker who released all the names and contact information of their customers. This was quite embarrassing for many well-known people whose names were found on the Ashley Madison data lists.

Sex worker sites

These sites cater to people who are looking for sex workers. Prostitutes, strippers, call girls and boys, escorts, and sex workers of all types advertise for customers on these websites. Men must sometimes pay to have full access to these sites, or the workers pay to advertise their services.

Obviously, a big risk on all these sites is that you may not get what you think you're getting. I have encountered many patients who have been physically injured when they went to meet someone they connected with on one of these sites, and they were hit over the head and had their laptop, wallet, and car stolen. There is also the risk of a disease if you're having unprotected sex with someone you don't know. In general, this is *not* safe behavior, but if done while in an addicted state, risks are often minimized or denied. There are other risks as well. You may be the victim of a data breach, where your name and private information are stolen and shared without your knowledge, and of course, many of the sexual services sold on the professional sites are illegal. Because prostitution is illegal in most states, a customer can be arrested if they are caught or if law enforcement happens to be operating a sting operation.

All these apps and sites can easily become addictive, and that is where Internet addiction, sex addiction, and porn addiction may converge and become a problem. My experience is that most porn addictions take a long time to develop, years in some cases. In other cases, they can escalate quickly and even progress to real-time sexual contact, including sexting, online sex, phone sex, sexual hook-up sites, prostitution, and other sexual behaviors. Some of my early research found that the more a person looks at pornography, the more likely they will progress to other forms of sexually addictive behavior, although this varies depending on the person's situation. Again, the ease of access, availability, and anonymity of the Internet are a perfect match for all types of sexual behavior, but especially pornography.

Pornography addiction can lead to other sexually addictive behaviors and can also lead to desensitization of the sexual response, leaving a man with premature erectile dysfunction. Sexual arousal and sexual function need neurogenic stimulation to work properly, and if you habituate to sexual images and videos, this arousal mechanism can be dampened and potentially damaged.

Webcams, toys, virtual reality, and robot sex

A few years back, I did a video television series called *Wired for Sex*. It was clearly ahead of its time, and it outlined all the various interfaces between sex, the Internet, and technology. Most of what I discussed on that video series in 2003 has come to pass. Sex and the Internet are now inexorably linked. In fact, you cannot even talk about pornography or pornography addiction without the Internet being part of the conversation. Here are a few examples today relating to what we talked about almost 20 years ago:

>> **Webcams:** You can put a video camera anywhere you want and broadcast sexual activities over the Internet, from dancing to stripping, to masturbation or hard-core sex for others to watch (and participate in). Webcam sites, as they are called, are generally paid for via subscription or tipping, much like going to a strip club, but in this case, everyone is in the privacy of their own home in front of a screen. The performers can see your typed messages (and reply to you) on their screens asking for specific sex acts and making requests, often accompanied by a monetary tip. Most of these transactions take place using various financial media, including Bitcoin, credit cards, and other electronic financial systems. When it comes to the adoption of cutting-edge technology, the sex industry is often at the forefront.

>> **Internet porn sex toys:** This is where Internet technology really meets sexuality. Pornographic videos and films have been around a long time, and the next level of Internet pornography links sex toys for men and women with a sex scene in a video or live sex act. The toy is connected to the Internet and is synced via a program to mimic the sexual acts of the onscreen performers. You can direct or change the action if multiple scenes are filmed. This takes on new meaning once CGI pornography really takes off, as it will be much less expensive to produce, and users will essentially be able to customize their own pornography — possibly with kinky and fetish requests that are specific to their preferences. In some cases, the sex toys are like life-like sex dolls and can be made to perform Internet-driven sexual actions that the user can literally have sex with. Other times the actor or webcam person is using the toy themselves and is taking requests (for tips), and the toy responds with escalating intensity, commensurate with tips.

>> **Virtual reality sex:** This is still coming along, and it is exactly what it sounds like. You are immersed in a virtual world via an avatar having sex with another avatar; the more complex issue that is not fully resolved is how they transfer the simulation effectively from the virtual icon to the physical body in real time. Although there are crude versions of this technology, it is not quite there yet.

There are possible implications for use in medicine in patients with spinal cord injuries and other persons with physical disabilities. A lot of good can come from the merging of sexual health and human sexuality with technology, and just like all powerful forces, there will be both productive and counterproductive uses.

>> **Sex robots:** Years back, I did a show on sex robots, but now the technology has really arrived. Pretty soon, and to some extent even now, users will be able to have sex with a robot partner. The sex robot can be programmed by the user or by the manufacturer, with varying levels of complexity and sophistication. There are cases where such sexual robots are being used in sex offender treatment to help offenders manage compulsive sexual behaviors. In this case, technology may be useful in treating compulsive or addictive sexual behavior. There are also potential applications for people with disabilities and those with geographical challenges.

Fantasy, role play, and anonymity

One of the major strengths of Internet sexuality and pornography is that it enables users to play out various sexual fantasies, either through viewing desirable pornography or connecting with like-minded people who want to experience similar sexual behaviors or fantasy. The Internet's perceived anonymity is very attractive for people who want to enact these fantasies and to engage in various role playing. Here they can take on a new sex role — a different gender, a type of sexual behavior they might not otherwise get to experience, or other fetishes. None of these fantasies or roles even need be acted out, but simply discussed with other like-minded people, in chat rooms, who want to engage in such discussion.

One interesting twist to this was a legal case I was asked to offer commentary on in an HBO documentary special titled *Thought Crimes: The Case of the Cannibal Cop*. This was a case of a New York City police officer and his arrest for engaging in online fantasy discussion that involved sexual torture and cannibalism. As distasteful (no pun intended) as this concept is, it brought to light interesting legal and psychiatric issues.

The case background is that his wife found the website where he was discussing his detailed desire to commit these heinous acts, some of which involved her. She was naturally frightened and called the police, and he was arrested. The case took many turns, and although initially convicted, he was later acquitted. He is currently free, although no longer a police officer. The issue here was that ultimately, no matter how objectionable, the court ruled that there was no proof that his fantasies could necessarily be predictive of actual intended behavior (the difficulty in predicting behavior is consistent with the state of the art in current psychological science). The case is currently under appeal to the U.S. Supreme Court, and the implications are highly relevant for many aspects of Internet use; this case directly speaks to the use of fantasy and discussion of that fantasy when it is violent and/ or objectionable.

If I could choose only one of the Internet's most powerful features when it comes to sex and pornography, it would have to be *anonymity*. It is the *perception* that what you do online cannot be seen or heard by another person, which makes it such a fertile ground for all types of sexual behaviors, and especially for pornography.

WARNING

The problem, however, is that the Internet is *not* actually anonymous at all. In some ways it is the least anonymous form of media or communication there is, because there is a potential record of everything you say and do online. Not that everyone always has access to all of this information, but I can assure you that if there was a legal reason to get your browsing records or web searches, they could be accessed. Take care if you think that what you do online is anonymous. Most people forget to even erase their own web history on their computer, phone, or tablet. So many of my patients are caught by loved ones simply looking at their devices. Care should be taken with what you say and do online.

REMEMBER

Everything you do online is forever.

Fetishes

Perhaps another strength of the Internet is its ability to connect like-minded people who share fetishes. One of the more therapeutic aspects of online sexuality is the ability for people to connect with those who may share their interest in fetish sexuality. Simply defined, a *fetish* is a nontraditional arousal pattern that is often required (or can be additive to other, more traditional forms of sexual arousal) to achieve sexual excitement. *Paraphilia* is the fancy term referring to a fetish.

There are many fetishes. The most common one is foot fetishes and BDSM, which stands for *bondage, discipline,* and *sadomasochism.* Sounds funky, but it is essentially where one person likes being tied up or bossed around sexually and the other person likes controlling and giving the orders. It works if you have the right partner, and many times people write, speak, or fantasize about the fetish without ever physically acting it out.

WARNING

Some fetishes can be problematic if they involve unwilling or unknowing victims, like voyeurism (where you like to watch people without them knowing or consenting). Some fetishes, if acted out, can carry serious legal consequences.

Many porn sites cater to specific fetishes. There is pornography available even on free sites like YouPorn, Hamster, and Pornhub for foot fetishes, BDSM, fat fetishes, thin fetishes, old and young fetishes, gender play, and much more. Most of this content is poorly depicted in the videos but serves as a general theme for specific arousal. Dozens of various sexual themes and preferences are organized on dozens of Internet porn sites.

The Psychological, Biological, and Legal Repercussions of Online Sexuality

Online sexual behavior now represents a large area of human sexuality. The Internet is used not only for viewing, sharing, and exchanging pornography but also for webcamming, live sex chat, meeting people and hooking up, sex education and sexual exploration, fetishism and paraphilias, gender and sexual orientation play, and experimenting with role playing and taking on alter egos. Some of these areas can cross the line online and lead to legal, psychological, or medical repercussions, so care needs to be observed when mixing sex and the Internet.

Appreciating the Internet as a place to learn and teach about sex and sexuality

As a teaching tool, the Internet is a great asset — there is much to learn online. Although the Internet in no way takes the place of real-time learning, it can be extremely useful in augmenting learning. When it comes specifically to sexuality, however, the Internet can be a resource for information (some good and some not so good) about human sexuality, medical and technical aspects of sexuality, and general information about the topic. Forty percent percent of kids don't receive any formal sex education in school, and many parents either do not or cannot provide the information — or they lack the comfort level to discuss the subject. Positive or negative, the Internet can fill the information gap, and in many instances, people can learn about aspects of sexuality from credible online resources.

I am not suggesting that pornography is a great resource for sex education (as there are many negative messages found in pornography that distort realistic human sexuality), but there are many *other* viable resources online. In many cases, however, porn may be the only option for finding out about sex in a world with a duality when it comes to sexuality. In this country we celebrate sexuality: We advertise with it, market with it, and inject it into our art, films, television, and all forms of media, yet at the same time, we have a somewhat puritanical and Victorian view of sex and sexuality. This duality pushes healthy sexuality under the carpet as there is still a covert undercurrent of discomfort in dealing with the subject. This schism between *overt* and *covert* sexuality can lead to ignorance and unhealthy sexual behavior.

There are so many aspects to sexual behavior and human sexuality that even as a sexual medicine expert, I cannot keep up with all the new developments that involve sex and the Internet. The Internet is a potential resource (provided the sources are vetted) to be used by everyone who has an interest in learning about their own sexuality and understanding normal and pathological sexual behavior.

The Internet — including some social media, chat rooms, and specific apps and sites such as Reddit, Wikipedia, Facebook, Twitter, YouTube, and others — offers potential online resources that are sources of information, support, and social connection regarding sex and sexuality. The American Association of Sexuality Educators, Counselors, and Therapists (AASECT) and the Sexual Medicine Society of North America (SMSNA) are good resources for information. In today's complex sexual world, there are many more options regarding sexual identity, sexuality and gender support groups, gender identity issues, and sexual behavior than at any other time in history. There are also more socially accepted options for expression of new ideas in gender definition and fluidity, cisgender issues, sexual preferences, and alternative sexual and nonsexual lifestyles than ever before — indeed, it's an exciting time in human sexuality, and the Internet is a major contributor to these advances. The Internet can indeed be a place to find information and support in ways not before available.

Weighing the social, moral, political, and ethical impacts

Sexuality and pornography have many social, ethical, moral, and religious implications. Part of my approach as a medical professional is to attempt to look at the science and medical facts regarding sexual behavior, including addiction to sexual content and online sexual behavior. I am not ignoring values-based approaches, but I generally try to stick with biological and psychological facts.

Although everyone is entitled to their personal values and beliefs, it is my opinion that those beliefs should at least be informed by science. We could certainly debate the moral benefits or deficits regarding pornography or sexual addiction issues, but to some extent, that ignores the bigger question of helping people deal with situations as they are and treating those people who have a problem with their addictive sexual behavior, or any issue regarding sex and sexuality for that matter.

REMEMBER

You should not make political decisions regarding human sexual behavior. In my opinion, sexual and gender identity, sexual orientation, and sexual behavior cannot be legislated or controlled by social, religious, or political actions. I understand that this view is perhaps controversial, but the data is very clear that sexuality and sexual behavior are inherent and not responsive to external control. That is not to say that someone suffering from sex or porn addiction should not be treated, or that people who are struggling with gender identity or transition concerns should not helped or supported. Sexual identity, sexual behavior, and addiction are biological and psychological issues, not moral ones.

REMEMBER

It is important to be mindful to isolate the social, political, and moral issues associated with people dealing with sex, sexuality, and sexual issues. Such individuals are already dealing with shame, stigma, and bias and do not need more to deal with while they struggle with their problems. The most important thing we can do is to be nonjudgmental and compassionate, even if the issue is foreign or unknown to us or makes us uncomfortable. However uncomfortable we are, the person struggling with a porn or sex addiction or other sexual concerns needs human understanding and support, as well as possible professional help.

Exploring potential health and medical consequences

REMEMBER

Addiction is *never* a healthy choice. The fact is that addiction is *never a choice at all.* No one grows up wanting to be an addict of any kind, and certainly no one aspires to become an Internet porn addict, or to have their sexual behavior get out of control. People have problems. People have illnesses, and when they are ill, they need compassionate help, not judgment.

All addictions are essentially bad for your health. Addictive behavior limits healthy choices and fosters unhealthy ones. Porn addiction, like all addictions, increases stress and cortisol levels, reduces real-time sexual behavior, and can lead to erectile dysfunction, poor sleep, damaged relationships, loss of one's job, expulsion from school, and potential legal consequences if illegal pornography is involved. Generally, all addictive and compulsive sexual behavior takes you down paths that can lead you to being physically hurt. If your sexual behavior escalates to real-time contact, then the health and personal risks can escalate even further to include venereal diseases, AIDS, physical violence, loss of marriage and relationships, and financial impact.

There is much confusion when it comes to the interaction of pornography use with real-time sexual relationships, but there are some potential positives here. When used in moderation, pornography can be a stimulating form of content that can provide a consenting couple or individual with new ideas, fetish stimulation, creative positions, and simply added sexual arousal, which is perhaps what pornography does best. There are perhaps appropriate uses of moderate amounts of pornography for individuals who have sexual arousal disorders and need greater external stimulation. Lastly, for those people who do not have or cannot have a real-time sexual relationship (due to physical or emotional disabilities or geographic limitations), pornography can provide a safe alternative for arousal and masturbation.

Considering legal issues and risks

WARNING

As I mention throughout this chapter, there are some legal risks when it comes to pornography and real-time sexual behavior. The most notable risk that I mention is the gravitation toward illegal forms of pornography after becoming desensitized from viewing excessive amounts of traditional pornography. Typically, this may involve the illegal use, downloading, or transmission of child pornography; sometimes this illegal use is facilitated by logging onto other people's computers using file-sharing software to view pornography of unknown types. This can be quite dangerous, as you do not know what they may have on their computer when you download content or whether the computer you are logging onto is run by a government or law enforcement agency.

Addictive sexual behavior often blocks a person's logical judgment and reasoning ability. Many people make exquisitely poor decisions when it comes to their compulsive sexual behavior; porn use, acting out sexually with affairs, using webcam sites, visiting sex workers, and other behaviors that can have significant legal, health, and emotional consequences. There have been thousands of cases where people's careers, marriages, livelihoods, and freedom were brought down by addictive or compulsive sexual behavior. Sex is not something to be feared, but people need to be conscious of the powerful neurochemical effects it can have on their brains. Sexual arousal is *designed* to block logic and your better judgments because at one time, it was to our species' evolutionary advantage to do so — this is no longer the case.

REMEMBER

When it comes to sexual behavior, whether looking at pornography, visiting a sex worker, having an affair, or going on sexual dating sites, the anticipatory arousal may activate dopamine pathways that begin to erode logical thought and reasoning. Although I'm not attempting to remove personal responsibility, if the sexual behavior is part of a *progressive addiction*, then this may remove some of the volition and choice. The best way to avoid the risks associated with these poor choices is to have the underlying sexual addiction pattern treated. Addicts generally do not make great decisions, so the best way to make better decisions is to treat the addiction and co-occurring psychiatric issues.

Discussing thought crimes: Crossing the line online

Earlier in this chapter, I mention the *Thought Crimes* HBO story that I appeared in. This case really brings up a much bigger question: How do we separate thought (including fantasy) from behavior? Not a simple task. In the HBO story, this person wrote out rather detailed and, if enacted, deadly behaviors and shared them on an Internet chat room that dealt with sadomasochistic sexual fantasies, including torture, rape, and cannibalism. Now there are many ways to examine this, and I am in no way excusing his behavior or such websites, but it raises an important point: *Where is the line online?* Can you express your deepest sexual thoughts and fantasies without fear that you will be punished legally or otherwise?

I would be remiss if I did not acknowledge that there are sexual behaviors and deviances that represent severe mental illness. The interplay between sex, psychopathology, and violence is real, and I am in no way attempting to minimize it. It is dangerous, and people have been hurt and killed by such mentally ill individuals. Rape is an example of a sexual crime that integrates violence with sexuality in an unhealthy and damaging way. In a less extreme example, there are consenting adults who engage in a degree of physical pain, dominance, submission, and bondage, and no harm occurs. I am not equating these extremes, but simply pointing to a continuum on which sexual behavior and sexual arousal lies — and there isn't always a clear line.

What is okay if it's between consenting adults? That is a great question. Unfortunately, it is not a question I have a simple answer to. For the most part, sexual behavior that is consensual and accepted by all involved parties is fair game. The problem occurs when someone consents when they should not, or they consent when they are legally underage, or they consent but they are mentally ill in some way, or if they consent when they are ambivalent but want to please someone else, and lastly if they consent and change their mind. It is complicated to say the least, and human sexuality, both healthy and addictive, encompasses a very wide range. Care must be taken in understanding all the subtleties and variables and helping those who need and want help.

3

Diagnosing and Treating Internet Addiction

Discover how to assess some of the key signs and symptoms of Internet and technology addiction. You examine the cognitive, emotional, psychological, and medical impacts of excessive screen use and Internet addiction.

Find out several self-help strategies for the addict, family members, and loved ones. Healing and recovery from any illness or life challenge should involve self-help and support of various types. Self-help strategies can be useful to the addict and their family in both helping to achieve change and managing when a loved one is not motivated to change.

Experience new digital habits while keeping in mind the human element of the Internet. The idea is to bring greater mindfulness and balance to your screen use and remember that the Internet is a tool, not a destination. Finally, appreciate living in a real-time, social world by beginning to overcome your urges and triggers, and learning to manage boredom intolerance.

Look at various treatment options, including parent consultations and coaching, outpatient treatment (in person or via telemedicine), intensive outpatient programs (one- to four-week options), day treatment/partial hospital, and residential treatment options. Sometimes, a mixture of treatment options may be utilized, as well as addressing co-occurring psychiatric symptoms and disorders; treatment may also include IT monitoring, blocking, or limits.

Chapter **10**

Identifying the Signs and Symptoms of Internet Addiction or Overuse

A ll addictions create behavioral changes, with the ultimate symptom being narrowing your life focus and limiting your real-time living behaviors. The signs and symptoms of Internet and technology addictions are not dissimilar to those of other addictions; however, some aspects of Internet addiction are specific to the overuse of the Internet and screens.

Internet and screen addictions are relatively new areas of study; as a result, addiction medicine has not yet fully developed definitive diagnostic markers for these problems. Currently, there are numerous diagnostic labels for this disorder, which include Internet use disorder, pathological or problem Internet use, and Internet addiction disorder, among others. Perhaps the most agreed-upon term to date, which is specific only to video game addiction, is *Internet gaming disorder,* which appears in the *Diagnostic and Statistical Manual of Mental Disorders,* 5th Edition (DSM-5) as a provisional diagnosis. It also appears in the World Health Organization's (WHO) International Classification of Diseases (ICD-11) and is listed as *Gaming Disorder.*

Currently, no final consensus exists on which diagnostic markers and symptoms are necessary to warrant an Internet and screen addiction diagnostic label. However, there is general agreement on many of the relevant issues presented in this chapter, and there is an overall agreement that a clinical problem exists with compulsive Internet and screen use. Going forward, it's likely that there will ultimately be several subtypes of Internet and media-based addictions, although the final nature of those nuances is a work in progress.

REMEMBER

Addiction to the Internet isn't based *solely* on the device or the content. Rather, the *Internet modality* itself is inherently addictive, along with the interaction of the consumed content.

The type of screen device you use to access the Internet is secondary to the addictive experience. Perhaps the one exception to this is the smartphone, which has an added feature that, in my opinion, increases its addictive potential: notifications. Notifications are invitations to your nervous system to come and see what is waiting for you; sometimes what you find will be good, sometimes not, but the result is that you picked up your phone and your eyes are on the screen. In fact, the dopamine release for an anticipated event is twice that of the actual rewarding event; a notification serves as a potent trigger to this innervating reward — this is the *slot machine* or *maybe* effect that I talk about throughout this book (and introduce in Chapter 4).

Recognizing Cognitive Symptoms

Perhaps the most pervasive (as well as impactful) aspects of Internet addiction and overuse are cognitive changes. Some of these changes are likely related to brain-based alterations, and others appear to be a result of psychological and cognitive factors that inform the addiction experience. We see similar cognitive impacts in all addictions, irrespective of drug or behavior of choice. Denial, distortion, rationalization, minimization, and justification are some of the most common cognitive changes we see with addiction — these mechanisms help maintain the addiction, and can reduce openness and motivation for treatment.

Denying and distorting the need to be online

Rationalization, denial, and distortion serve as essential defenses for your psychological and cognitive functioning. These cognitive processes are actively (although often unconsciously) used to manage incompatible beliefs or to avoid addressing your behavior when it is problematic. They are also important when managing your *cognitive dissonance;* when you have competing ideas but may be committed to one set of beliefs or behaviors, you would then need to manage the competing beliefs to keep doing what you want to do and to function efficiently. This is an active cognitive process, where you must rationalize, deny, or even distort reality to maintain your current thinking or behavior pattern.

THE BRAIN, VIDEO GAMES, AND EMOTIONS

First, regarding video game violence, evidence shows that areas of the brain responsible for response inhibition and modulation are impacted by the brain's need to *inhibit normal emotional responses* when viewing upsetting or violent video games and other violent video content. This makes sense, in that a person would need to inhibit their emotional responses to interact with such content; the problem occurs if this pattern of response suppression is maintained over the long term and becomes the habitual and go-to response for managing emotions. What many people don't know is that the brain is very involved in regulation of emotion, and certain emotions are healthy to feel and express in a regulated manner. Therefore, the idea that excessive video game use might serve to repress emotional expression and emotional regulation is noteworthy.

Another area of neurological impact is the *Go* network (which is related to the brain's ability to start and stop tasks) and the RaS (the RaS, or reticular activating system, is related to arousal and neural activation). Processing in the brain that is involved in the starting and stopping of tasks seems to be impacted by high amounts of video gaming and Internet screen use. These brain functions are also involved with impulsivity and managing distraction; there are various experimental ways of using neuroimaging studies to examine tasks using a go/no-go format, and evidence suggests that there may be issues with gratification delay and impulsivity. This dovetails with other research on suppression of pre-frontal cortical inhibitions seen in many addictive disorders.

It's interesting to note here that we typically see a very high co-occurrence of ADHD with patients who are addicted to the Internet and video games. It is my assumption that part of the reason for this is the *stimulus blunting* and *suppression* seen in ADHD, and the fact that Internet and video game use may override this stimulatory threshold. Both the Internet and (especially) video games are highly stimulating on numerous levels. In other words, the noise level is tuned so high in an ADHD brain that the desired signal (what you want to pay attention to) may be too low. The use of video games, and many Internet behaviors, may amp up that signal enough to punch through this attentional threshold.

To summarize in a less technical way, video game and Internet overuse desensitizes people to general life stimuli and reduces their ability to express emotions. Video game and Internet addictions are stimulating enough to affect the emotions of heavy users and addicts with ADHD, and to override any attentional deficiency.

The following sections cover denial and distortion; I cover rationalization later in this chapter.

Denial: Not just the longest river in Egypt

REMEMBER

Denial is a defense mechanism, and it does exactly what it's supposed to do. Denial allows you to engage in a behavior and to ignore (deny) certain consequences. To a large extent, this requires some cognitive brain power as it takes some effort to ignore aspects of reality that may be staring you in the face. The amazing thing about denial is the extent to which a person can ignore such circumstances when everyone around them sees the reality of the situation.

A case in point might be a typical set of circumstances that we frequently see at our clinic. The addict is ignoring their family, doing poorly in school, and perhaps not taking care of themself, but when asked if they are aware of the negative impact of their screen use, they will say there is *no negative impact.* Or they will use another defense and *rationalize* why they are having all of these consequences (and typically it is *not* because of their screen use — I discuss rationalization later in this chapter).

Cognitive distortion

REMEMBER

Distortion is related to denial and rationalization, and it's used when a person wants to reinterpret reality in a different way. We all do this occasionally to some extent. But when an Internet addict does it, they will literally miss the reality of their situation and view all the consequences of what they have been doing from a distorted perspective. In other words, they change the reality of the actual event on either a *causal* level or a *consequential* one. What this means is that the addict distorts the *cause* of their addictive behavior, and they also distort the *resulting* negative impact. This defense is more conscious and requires some semi-conscious brain power to facilitate.

One key area of distortion occurs around the time a person spends online. My original research in the late 1990s found that people experience a high degree of *time distortion* when they are online; time is hard to keep track of when you're on a screen connected to the Internet. Most of us feel this when we are online, and for that reason, we know the Internet can act like an intoxicating substance — altering the perception of time, space, and emotion.

One of the first things parents and loved ones tell me in a consultation is how much time their child is spending online. This is, of course, not surprising because it's difficult to develop problems with online use to the point of interfering with your life, unless you are *dedicating a lot of time to it.* Most patients (and many users of the Internet as well) distort the amount of time they spend online to a drastic extent. Whereas they might be on for 6, 10, 14, or 18 hours a day, they may only recognize being on for a few hours. Often this discrepancy is the source of many family disagreements. This is one reason we often use software that monitors the time spent online and what sites are visited so that we have real data to make treatment decisions with. The amazing thing about distortion, however, is that even when confronted with this use data, our patients still say it is inaccurate and argue for their distorted perception.

REMEMBER

It's important to keep in mind that cognitive distortion is not entirely volitional and may not be an act of willful defense. Although there can be some opposition-ality and willfulness, it's a neurocognitive process that contributes to such distortion.

But everyone's on the Internet! Rationalization

Rationalization is the ability to justify your behavior with the excuse that everyone else is doing the same thing. I'm reminded of when I was a child and I wanted to argue in favor of something my parents were against, and my go-to statement was that *everyone else was doing it.* Now, I really had no idea if that was, in fact, true, but it seemed like a prudent approach at the time to get what I wanted.

REMEMBER

Yes, everyone does use the Internet (well, not everyone, but close), but that has nothing to do with the fact that *some* people can and do become addicted.

The Internet is not a drug; it is a useful tool for communication, information, commerce, and entertainment. So obviously it's a good thing, right? That was the prevailing attitude back in the mid-1990s when I started researching and writing about Internet addiction. This was a time that the Internet could do no wrong, and there was an all-out stock boom with Internet technologies. It was going to be the answer to many social issues and serve as a *digital leveling field,* whereby everyone would have access to all the same information. So, I presented a paper titled "The Nature of Internet Addiction: Psychological Factors in Compulsive Internet Use" at the 1999 convention of the American Psychological Association. This paper, which was later published and became the basis for a book, was one of the first studies to argue that the Internet was, in fact, addictive. Of course, it wasn't addictive for everyone, but a sizable number of users (5.9 percent) met the criteria for addiction, and an even larger number overused the technology. Subsequent studies have found similar percentages of addiction to the Internet.

The dominant popular thinking back then was to question how the Internet could be addictive. My later work and the research of many others demonstrated that the Internet was not just a tool, but also served as a *place to go* and as a *process* that acted like other mood-altering and addictive substances and behaviors, and neurobiological evidence strongly suggests that the Internet seems to impact the brain's reward centers, just as other addictions do (see Chapter 2). In short, the Internet, for all its good and all its functionality, is a *digital drug*.

A new normal

There is no question that the Internet is now part of everyday life. It is here to stay and will be used by generations to come in all its new iterations. The popularity of a drug or behavior like the Internet does not prevent or justify addiction to it. Let's not blame the screen-addicted victim either. People have problems and sometimes those problems involve addiction to popular substances or behaviors like the Internet.

Alcohol is certainly part of our daily culture; so, too, are certain substances like marijuana, as well as gambling and sex. Many people engage in these and other addictive substances and behaviors with few or no problems. In fact, most people (estimated at upwards of 90 percent) can engage in these activities and recreational substances without developing a problem. Unfortunately, the fact that so many people *don't* develop a problem does not take away from the fact that a sizable minority of them *do* develop a problem and that sometimes this problem develops into an addiction.

It's all in the game

Many parents and loved ones are sympathetic at first (often for many years) to the normative aspects of Internet and video game use. In fact, most of the devices, screens, software, apps, games, and high-speed Internet access are supplied by someone. I cannot tell you how many high-end, custom gaming computers costing thousands of dollars have been supplied to a budding Internet or video game addict, not out of malice, obviously, but out of love and for the fact that it seemed so harmless. After all, many kids have a Nintendo Switch, or PlayStation, or XBox, or gaming PC, as well as a smartphone. There is an often a simplified attitude that children, adolescents, and young adults are *just going online* or *just gaming.* And as the word *game* implies, it's for fun and entertainment, like Monopoly or checkers.

REMEMBER

But the fact remains: I've been in the addictions field for well over 30 years, and I have never treated anyone for a Monopoly or checkers addiction. Never. The fact that most people can engage in behaviors that some people get addicted to makes these behaviors no less addictive.

So, even if "everyone" is on the Internet, some people still *do* get addicted, but those folks who are addicted may rationalize that they *don't* have a problem because everyone else is doing it. Rationalization is often based on comparison or justification of what other people are doing without really knowing how it is impacting others.

Studying Psychological and Emotional Symptoms

As you find out in this section, many psychological, emotional, and psychiatric symptoms can and do arise from excessive Internet and screen use. There are also many co-occurring psychiatric problems that are either caused or exacerbated by heavy screen use. And there are still other diagnoses that probably existed, to some degree, prior to the development of screen-related issues and provided fertile ground to assist in the development of a larger problem.

A large majority of my patients seem to have neuroatypical brains; the most frequently seen issues are ADHD, learning disorders, sensory issues, and, to a lesser but still significant extent, ASD (Autism Spectrum Disorder). Other preexisting conditions that I've seen frequently are social avoidance, social anxiety, generalized anxiety, and depression.

REMEMBER

It's important to note that sometimes it's difficult to discern which is the chicken and which is the egg. Human behavior, including addictions, is not directly based on cause and effect. Often, multiple factors interact, including genetics, environment, and some of the other circumstances I've noted. It's likely that Internet addiction is more complex in its etiology, and part of helping the addict is to help them identify some of these contributory factors.

REMEMBER

It's important to note that although it's difficult, it is necessary to help transfer the positive psychological and neurochemical benefits from a person's screen use to real-time living. This is a gradual process and cannot be attempted half-heartedly or inconsistently.

Anger issues

So, what are the personality and behavioral characteristics that change because of Internet and screen addiction? Several key changes deserve to be noted here; these include changes in the addict's withdrawal from previous real-time living behaviors and social activity, as well as reduced productivity. There is also evidence to suggest that changes may occur in empathy and social perspective, as well as increased attempts at multitasking, despite its poor outcomes.

There is unlikely to be an issue with excessive Internet and screen use unless notable increases occur in the amount of time spent online; there will always be significant increases, although they may occur gradually over a long period of time.

The addict may experience increased anger, aggression, and hostility related to any attempts to limit, remove, or control their Internet or screen use. This is especially notable with video gaming and smartphones, where I've seen significant anger and family discord. This can also be part of a long cycle of parents removing technology out of frustration or concern, only to return it after the addict has an angry outburst or threatens to harm themself. Sometimes the addict can be quite threatening or aggressive to their parents, and the police may need to be summoned when serious threats or violence occurs.

It's hard for a parent to know how to manage these situations, but my professional opinion is that *safety always comes first,* and that removal of technology must be part of an organized and structured treatment plan; just removing the tech does nothing without the context of a comprehensive treatment plan. I never recommend escalating conflict without a plan, and if the addict becomes abusive, threatening, or destructive, it's always safer to call the police to assist. It is never a good idea to spontaneously facilitate this level of anger and aggression, because there is a chance that something dangerous could occur — an organized clinical treatment plan is needed to manage these issues.

Mood and other psychiatric symptoms

The most notable psychiatric symptoms that I see are related to depression and anxiety. Sometimes depression is secondary to long-term excessive Internet and video game use and at least may be partially due to reward deficiency syndrome, whereby normal stimulation and regulation of mood becomes skewed due to excessive screen use (see Chapter 2). My experience supports that many, if not most, of the patients I've seen have had anxiety issues prior to their Internet addiction, but these symptoms often get worse or at least covered up while people are on a screen; many times, social anxiety is also present, and screens allow some type of social interaction at a safe distance.

Life withdrawal

Withdrawal from real life is a common symptom in all Internet-based addictions, and it is difficult to develop an Internet addiction without withdrawal (in some ways) from real-time living. This includes friends, family, work, school, and self-care. Do not be fooled by the addict's insistence that they have intimate and rewarding relationships online and that their gaming and social media friends are all the friends and relationships they need. The level of friendship found online may have many positive features, but it is not always an equivalent substitute for real-time connections and relationships.

In my clinical practice, I am reminded of the preceding fact daily; I often see patients via telemedicine, and this virtual option has been invaluable, especially during the COVID pandemic. It also allows me to do consultations and provide parent coaching at locations too far for people to drive to see me. However, it is clearly not equivalent. Psychiatry is a great field to utilize telemedicine, but it does not provide the nuanced cues and information needed to *feel* the patient in a complete way. Some of this is difficult to *quantify,* but the *quality* is real. It is the same for friendships and relationships. Social media updates and postings, texts, Snapchat, Instagram, TikTok, Facebook, Google Chat, messaging (instant and direct), and Discord are all good approximations for human connection, but not quite the same. My thoughts are that such digital methods may serve as communication shorthand, while removing much of the nuanced quality that provides the richness of human interaction.

REMEMBER

A major symptom we see with heavy Internet use is the avoidance of real-time living and all the skills and tasks associated with being a child, teen, tween, young adult, or even an adult. We often see a significant delay in the developmental level of age-appropriate living skills, and the more these skills are avoided, the harder it is to address them. Failure to launch is a frequent issue we see.

Adulting, as we call it, is hard and requires that a lot of mundane and boring tasks be done — and that means some attention and gratification delay. A big part of treatment is to help facilitate the skills and emotional readiness to tackle age-appropriate behaviors of living in the real-time world, and to get back on a reasonable developmental trajectory.

Depression

I discuss depression in several places throughout this book, but it's perhaps one of the most notable results of excess screen use. It is also important to note that many preexisting and co-occurring psychiatric issues are seen with Internet use, and as mentioned earlier, some are contributory to Internet addiction, such as ADHD, and some may be resultant, or both. Depression (which may in part be related to the reward deficiency syndrome that I cover later in this chapter) is a commonly seen feature. Many parents and loved ones describe a significant personality and mood shift, where they may feel as though they have lost connection to the addict. Often, what they are describing is depression in some form, as well as the addict's disconnection from real-time living. One of the saddest parts of treating this issue is hearing how parents feel they have lost their child, but after treatment, they feel they have their child back.

REMEMBER

It's important to clarify that depression is a complex disorder with many types and subtypes, some of which are more serious and debilitating than others. It is beyond the scope of this book to detail all the different types of depression. Instead, I will outline some of the more common features that are observable. Like all symptoms related to Internet addiction, treatment should be focused on *both* the addiction and the psychiatric issues. Focusing on one symptom alone is less desirable and is likely to be less successful. That said, I want to emphasize (which I do throughout this book) that addressing the psychiatric issues *without* managing the addiction will probably be less successful regarding treating the Internet addiction problem.

REMEMBER

Depression is marked by numerous symptoms, but the most common are lack of pleasure in previously pleasurable activities, low energy, sullen mood, lack of motivation, hopelessness, helplessness, reduced attention and concentration, possible suicidal or self-harming thoughts or actions, lowered sex drive, increased or decreased appetite, and disruption of sleep. Not every person has every symptom, and having some or many of these symptoms may or may not indicate depression. The only sure means to diagnose depression is to be evaluated by a mental health professional.

Depression can be a serious illness, and although it may, in part, be secondary to the addiction, it may require attention if you're seeking treatment for Internet addiction. Typically, a well-trained professional can address both issues, although sometimes multiple clinicians may be consulted, especially when medication may be indicated.

REMEMBER

A brief word here about down-regulation is warranted. Some of the depressive symptoms seen in Internet addiction are from the down-regulation of dopamine in the brain. It's impossible to accurately determine how much is due to what, but excessive screen use can lead to such down-regulation, which leaves the real world looking and feeling flat. The symptoms of this condition (reward deficiency) can, and do, look a lot like depression, but will usually improve if the addiction is treated and there is a return to real-time living.

Anxiety

It is said that depression involves obsessing about the past and anxiety involves worrying about the future; although there is truth to this, I am afraid it is not that simple. Anxiety, like depression (covered in the previous section), is a complex set of disorders and diagnoses, some of which are related to screen use; some may preexist heavy screen use and, in fact, may be an initial contributory factor in developing excessive screen use.

Social anxiety is perhaps the most common form of anxiety that I see with patients. Typically, although not always, there is a prior history of social anxiety or socially avoidant behaviors, but this only worsens with increased screen time. Some social anxiety and avoidance may be based on lack of skills, and sometimes it may be related to other problems such as ADHD or ASD.

Many times, Internet use is a means to manage or escape social anxiety with an easy, well-defined, and less threatening form of social connection. The Internet provides a buffer for the screen addict that allows for a titrated degree of social interaction. The problem is that it doesn't transfer well to real-time social skills and, in my experience, is not nearly as potent. It has its place, but it is likely not enough on its own — although the Internet addict may fiercely argue against this point.

REMEMBER

Anxiety, like depression, is a multifaceted and complex set of diagnoses that should also be evaluated by a trained mental health professional. The problem with anxiety is that the more time a person spends on the Internet, the less comfortable they feel with real-time social interaction; social comfort is a skills-based activity, and to some extent it is a *use it or lose it* phenomenon. Practice and rehearsal are required to feel comfortable and competent in a social world.

Anxiety, like depression, is very treatable, and most likely will be targeted as a treatment issue in Internet addiction treatment. Either way, anxiety is generally very responsive to psychotherapy, especially cognitive behavioral modalities, and can also be addressed with certain medications. Some of the medication we use can address both anxiety and depression symptoms. Anxiety, and especially social anxiety, can also be addressed using eye movement desensitization and reprocessing (EMDR), which is also used to manage addiction triggers, urges, and cravings, generalized and social anxiety, and trauma found in Internet addiction patients (as well as other patients). EMDR is an exciting addition to available psychiatric treatment options; it's a neurophysiological approach using bilateral stimulation of the brain utilizing rapid eye movement as well as bilateral audio and tactile stimulation.

Reward deficiency syndrome

Perhaps one of the most common symptoms I encounter is reward deficiency syndrome, which appears to be a direct result of the addictive Internet or video gaming pattern. As noted in Chapter 2 and elsewhere in this book, reward deficiency is seen in most addictive disorders due to the disruption of the regulation of dopamine and dopamine receptors in the brain.

The symptoms are profound, however, in that most previously pleasurable behaviors hold little interest to the addict. These can include social and family relationships, academics, work, hobbies, appearance and self-care, physical activity, and even sleep. It looks like depression (covered earlier in this chapter) but with some notable differences. In many cases, my patients have been previously treated for depression with both medication and psychotherapy, but with little success until we treated their Internet and video game addiction. The same holds true for prior treatment of ADHD, which we frequently see in these individuals.

REMEMBER

To treat reward deficiency, the underlying addictive behavior pattern must be detoxed and disrupted to allow changes to occur in these brain reward circuits. Most of the patients who come to me have not had an adequate opportunity to address the addiction in a concerted way, even though many of their psychiatric issues have often been treated. To counteract reward deficiency, after a period of detox and very limited use, it may then be possible to develop *positive reward transfer* to engage in more real-time living.

In my experience, to address reward deficiency, it's necessary to have a period of full or partial detox. This allows the brain to reset the neuropathways and post-synaptic receptors to more functional levels. Dopamine can return to pre-addictive levels, allowing real life to potentially take on more intrinsic value. If the brain develops up-regulated receptors in the reward center (as it does with an addiction), it will ultimately have down-regulated levels of available dopamine (see Chapter 2). The idea is to help more naturalistic aspects of reward be more obtainable, and to have less dependency on Internet and screen use for positive reward. I discuss, in greater detail, what a detox looks like in Chapter 13.

Positive reward transfer is exactly what it sounds like: It is the ability to transfer positive reward (and associated dopamine levels) to real-time living. This is not an easy task, and it is at least partially dependent on a detox period, followed by a focused, but limited, reintroduction to real-time living behaviors.

REMEMBER

Your loved one will likely *not* show any interest in alternate forms of pleasure and motivation if they have unfettered access to their addictive levels of screen use.

Seeing the Physical Symptoms and Health Effects of Too Much Tech

Numerous physical and health effects can manifest from too much screen tech; it's important to note that many of these symptoms affect some people, and other people have none. Some of the health problems reviewed in this section are rare, while others are more common.

REMEMBER

There is no way of knowing if you're experiencing symptoms related to Internet addiction without evaluating an individual situation. If you're concerned about your physical and medical symptoms, it's always a good idea to see a general healthcare practitioner to rule out other medical causes. I will often ask for a physical examination, lab work, or a consultation with their primary care provider if the patient has not seen their provider recently. Some of these physical and medical symptoms are logical and may be based on the unique factors of excessive time on screens but be unrelated to the type of content they consume online.

Sexual desensitization and erectile dysfunction

Perhaps one of the most troubling effects for my patients is the problem of erectile dysfunction (ED). Erectile dysfunction affects many men, particularly those over 40, and with greater incidence as they age. The symptoms vary, but essentially ED affects a man's capacity to achieve and maintain an erection sufficient to complete an act of intercourse. In more severe cases, ED can become insufficient for oral sex and masturbation as well.

Erection requires three aspects of biological functioning to work well. First, there needs to be enough of the sex hormone testosterone, followed by adequate vascular and circulatory functioning, and lastly, psychoneurological stimuli to elicit and maintain arousal. Some might call this the *psychogenic* aspect of erectile functioning, but suffice it to say that this is where fantasy, pornography, or real-time sex play elicits an arousal response, thus supporting an erection. Typically, some mechanical stimulation might also be required to facilitate and maintain an adequate erection, especially as men age.

Most men who have ED tend to have some deficiency or issue on the circulatory front, as impaired peripheral blood flow inhibits good erectile functioning. Medications called phosphodiesterase (PDE5) inhibitors can assist in compensating for this issue. Drugs such as Viagra, Cialis, Levitra, and Stendra can help. Also, insufficient testosterone can be a limiting factor for both libido and arousal, as well as the mechanics of erectile functioning, and some men benefit from hormone replacement therapy of testosterone.

The psychogenic aspects of arousal can be impacted by becoming desensitized to arousal signals from excessive exposure and use of online pornography (discussed in Chapter 9). I've seen numerous cases of men in their 20s who had no other medical factors that could account for their ED, but because they were no longer able to experience the psychoneurological aspects of sexual arousal, they could not instigate an erectile response. This was very troubling for these men, and they were all able to trace their deteriorated sexual response to their long-standing use of online porn. Some research suggests that it is reversible, but it requires a significant period of abstinence from all forms of pornography to re-sensitize to real-time sexual arousal.

Obesity and sedentary behavior

There is some interesting evidence to suggest that the ever-increasing body weights among American youth, and possibly adults as well, may be correlated with the increasing adoption and use of screen technologies. This makes sense. Screens require almost no physical activity and burn almost no calories. The only thing you're moving is your hand and fingers, and even then, almost no effort is required. If you plot the adoption and growth of the Internet, and all things screen-based, on a graph alongside average weights in the United States, you will see an almost perfect correlation. It's important to note that this doesn't prove a causal relationship, but it does seem suspicious and certainly makes intuitive sense.

For every 10 percent increase in a country's spending on information and communications technology (the Internet), there is a 1 percent increase in obesity levels. A 2019 study published in the journal *Advances in Nutrition* found that heavy Internet users had a 47 percent greater likelihood of being overweight or obese. The systematic review and meta-analysis that was done indicated that Internet use was positively associated with increased odds of being overweight and obese. Again, this makes perfect sense — with screens, we are less physical and more sedentary, and therefore we are becoming fatter.

There is ample evidence to suggest that the less we move our bodies (and you do not need to move a whole lot when online), the more we may eat. This seems counterintuitive — you would think you would eat less, as you're burning fewer calories — but that isn't the way it works. The science here is beyond the scope of this book, but suffice it to say that sedentary behavior induces metabolic slowing, which in turn increases appetitive behavior. The other factor here is that sitting in front of a screen for hours on end facilitates mindless, high-caloric grazing.

The avatars and characters played on MMPORGs (massively multiplayer online role-playing games; see Chapter 7) and interactive games never gain weight. They never get sick (they do die sometimes, but only temporarily). This fantasy can permeate the perceptions of an addicted gamer to the point that they engage in less self-care, including exercise and nutritional eating. It also demonstrates an invulnerability in online characters that we as mere mortals do not enjoy.

DVTs and blood clots

Movement, or lack thereof, is responsible for some heavy gamers and screen addicts developing deep vein thrombosis (DVT) or blood clots. This is a relatively rare occurrence, but dozens of cases have been reported worldwide, including in the United States, with the majority being in the Far East. China and South Korea have some of the worst problems in the world when it comes to video game addiction, and have seen many sudden deaths with heavy gamers.

Again, although a relatively infrequent occurrence, there are dozens of cases where during addictive trance of gaming or screen use, some users will sit for days without bathing, eating, or drinking. This lack of movement, combined with not eating or drinking, may lead to dangerous arrhythmias that can be fatal. This has resulted in some deaths — again, mostly in the Far East, but in the United States as well.

Because some gamers sit for days with almost no movement, and without eating or drinking, they have little need to release in the bathroom. They can also sit for prolonged periods of time with little to no sleep. Self-care and personal hygiene can be affected when gamers are lost in cyberspace.

TECHNICAL
STUFF

I have been to huge video game and Internet facilities in China (called *Internet cafes* or *PC bangs* in South Korea) that had hundreds of carrels set up with high-speed computers and monitors. They had an eerie quality. No one was moving and no one noticed me. They were also dead quiet (except for the faint clicks of mouses and keyboards), and everyone was wearing headphones. This reminded me of a heroin shooting gallery, where people are oblivious to all else but their intoxication and their drug of choice. I could not easily tell if they were alive. In these places, teens and youth will sit for hours or days, gaming with no adult interference. Some places offer showers and food, but my experience was that they were not there for a hot meal or a shower. They were there to game, and game they did, for 10, 12, 18, 24 hours straight, or more, often ignoring all else in their lives.

Repetitive motion injuries and phantom vibrations

What are repetitive motion injuries? They are strains or tissue damage due to heavy and repeated use of a limb or joint. Our bodies are not meant to do the same thing repeatedly for an extended period. When it comes to Internet and screen addiction, the most common area affected is the hands and fingers, although neck and upper-body strains are not uncommon from craning your neck downward toward your computer or smartphone screen. Here are some examples of the injuries:

>> **Texting thumb:** Texting thumb is what it sounds like and is from excessive and repeated texting and use of a smartphone. Both thumbs can become irritated and strained — often to a point of significant pain. There are no real assists for texting thumb, other than less use and the application of ice or NSAIDs (Ibuprofen or Aleve).

>> **Wrist and hand injuries:** These are common injuries resulting from extended use of a mouse or keyboard on a computer. They result from overuse of the

hand and wrist, and again can be quite painful. This is often seen in heavily addicted screen users and gamers, where they will surf or play for hours on end, moving little except their hands and fingers. It is also a common workplace injury. Ergonomic adjustments and devices can be used to reduce the likelihood of this type of injury, as well as obviously less screen use.

» **Head and neck strain:** This may be the most common of all screen use injuries and is not limited to Internet and screen addicts. All people who use screens, including smartphones and computers, crane their neck and upper back downward to maintain the correct eye-to-screen angles. This is okay for short periods of time, but because people are on their devices so much of the time, this can strain muscles and tissues, which become elongated and/or shortened from this extended pose. With an addict, this problem can be amplified due to the large amounts of time they spend onscreen.

» **Phantom rings and vibrations:** These are a well-documented phenomenon where you experience your phone ringing or vibrating, even when it is not. This is likely due to your nervous system being so tuned to the expectation of your phone's ring that it is experiencing the vibration or ring even though it has not really occurred. Part of this is likely due to the anticipatory neuropsychological response that your phone creates, and this seems to trick your brain into experiencing physical and sensory responses.

Eye strain

Another feature of heavy screen use is eye strain. The longer you stare at a screen, the more susceptible you are to eye strain; some of the strain is from the back lighting and blue light emitted from LCD screens, and some is from tracking your eyes up and down and across the screen on a repetitive basis. This can leave your eyes feeling sore and dry.

TIP

Eye drops can help in some cases, but the best treatment is to take a break and rest your eyes, or even better, spend less time on your screen if possible. Obviously, for Internet screen addicts, less use is not so easily achieved.

Gray and white matter changes in the brain

There is perhaps no more concerning aspect of compulsive screen use than whether there is an impact on the brain. I am frequently asked this question, as well as whether the potential brain changes are permanent. The good news is that most of the neurological and neuropsychological changes in the brain appear to be temporary — assuming, of course, that the heavy screen use is interrupted and the overall pattern of use is changed.

Research suggests that there are indeed white and gray matter changes in your brain that are related to heavy Internet screen use (see the nearby sidebar "Brain science research" for details). These changes are in addition to the disruptions and changes in the brain's limbic reward circuits, which I cover extensively in Chapter 2.

TECHNICAL
STUFF

White matter essentially consists mostly of long-range myelinated axons that exist more in the inner layers of the brain containing neuronal interconnections. *Gray matter* is different from white matter, in that it largely contains lots of cell bodies and fewer myelinated axons. The outer layer of the brain therefore has a more grayish appearance. The color difference is related to the white matter having a lot of myelinated axons, and the myelin has more of a whitish hue.

REMEMBER

It's possible that many neurological changes seen with Internet addiction may be *both a cause and an effect* — so it's hard to determine which is the chicken or the egg. This is consistent with my clinical findings in that the predominance of my patients with Internet addiction had ADHD, ASD, anxiety, or other neuroatypical symptoms *before* the development of their addiction, but seem to have worsened symptoms post-Internet addiction diagnoses.

Hypertension

Another common physical effect of heavy screen use is hypertension. Numerous studies show elevated blood pressure from heavy use of the Internet, although the direct mechanism of action is not clear.

In a recent study of 14- to 17-year-old heavy Internet users (10 to 14 hours a day), about 20 percent had notably elevated blood pressure. The average use was about 25 hours a week. Other studies support this finding, and it perhaps suggests a physiological stress-response mechanism to heavy screen use. Frankly, 25 hours a week is mild compared to some of the use numbers that are seen with Internet-addicted patients at our clinic.

In a more controlled study in 2017, 140 individuals who identified themselves as having problematic Internet use displayed increases in heart rate and systolic blood pressure, as well as worsened mood (depressive symptoms) and increased anxiety, after discontinuation of a controlled Internet session. This study supports many of the issues I see with patients, in that excessive Internet use seems to lead to *increased levels of stress,* as evidenced by elevated heart rate and increased systolic blood pressure (see the next section). The psychiatric symptoms for depression and anxiety (covered earlier in this chapter) are also typical of what is seen with heavy and addictive levels of screen use. By the way, anyone frequently spending too much time on their screens will likely develop these same symptoms, although perhaps to a lesser extent.

BRAIN SCIENCE RESEARCH

For the purposes of this chapter, I offer some examples of research studies looking at the brain science involved in Internet addiction, along with some explanations. I review, as painlessly as possible, structure and functional brain centers, along with the correlated neurological changes seen with Internet addiction and, more broadly, surrounding some of the neurocognitive effects of excessive screen use. Many of the white matter changes we see in other addictions are also observed in the brains of Internet and video game addicts:

- In a 2010 article that appeared in *The American Journal of Drug and Alcohol Abuse*, Jon E. Grant et al. looked at patient history and the features of tolerance, withdrawal, co-morbidity, genetic contribution, neurobiological mechanisms, and treatment response. They found a strong suggestion that behavioral (process) addictions, such as Internet addiction, do resemble substance-based addictions and that excessive Internet overuse is indeed an addiction.

- A study published in 2011 by Guangheng Dong et al. in the *Journal of Psychiatric Research* called Internet addiction the "world's fastest growing addiction." The authors found in an fMRI (functional magnetic resonance imaging) study that Internet addicts have increased activation in the orbitofrontal cortex in *gain* trials (money-like rewards in a game), and decreased anterior cingulate activation in *loss* trials, than do normal controls. The study suggests that Internet addicts have *enhanced reward sensitivity* and *decreased loss sensitivity* (something we see in other process or behavioral addictions like gambling). This produces a perfect storm for Internet addiction to develop, progress, and continue.

- A study published in 2012 by Guangheng Dong et al. in *Psychiatry Research* utilized an fMRI and a *Stroop Test* (a type of neuropsychological test that looks at the ability to inhibit cognitive interference) with young men with Internet addiction disorder (IAD). Results showed increased bold signal in anterior and posterior cingulate cortices compared with healthy peers. This finding suggests diminished efficiency of *response-inhibition processes* in Internet addiction. Response inhibition is essentially the *inability to stop, once started,* which we often see Internet and video game addicts having difficulty with.

- Daria J. Kuss and Mark D. Griffiths did an exhaustive review of the neuroimaging studies on Internet and video gaming addiction published in *Brain Sciences* in 2012. They reviewed 18 studies (some of which are mentioned in this list) and found compelling evidence for similarities with different types of addiction, notably substance-based and behavioral addictions such as Internet and video game addiction. They hypothesize that on a molecular level, Internet addiction is characterized by an overall *reward deficiency* that involves decreased dopaminergic activity (down-regulation). I discuss the concept of reward deficiency in Chapter 2 and how the brain adapts, thus making it less likely to find pleasure from real-time living.

- A study published in the *Journal of Psychiatric Research* in 2012 by Guangheng Dong et al. looked specifically at Internet gaming addiction (IGA). The authors observed that *diminished white matter integrity,* found with other addictive disorders, is also present in video game addiction, and this lack of integrity is associated with addiction severity, treatment response, and other cognitive impairments. Internet gaming addict subjects showed changes in the thalamus and left posterior cingulate cortex as compared to controls. Higher FA (fractional anisotropy) in the thalamus was associated with greater severity of Internet addiction and may suggest a *pre-existing vulnerability factor* for Internet and video gaming disorder and/or be a result of the *reduced neuroplasticity changes* from Internet and video game addiction. Reduced brain flexibility and adaptation is one thing we tend to see with all addictions.

  Internet addiction severity scores were also positively correlated with white matter integrity in only the thalamus, which suggests *greater reward processing* with screen-based, goal-directed behavior, and this is consistent with video gaming behavior and related dopamine changes.

- In a 2019 study published in *Scientific Reports,* Gergely Darnai et al. found a common brain-based feature of addictions in the form of *altered function of cortical brain networks.* Evidence suggests that Internet-related addictions are also associated with breakdown of *functional brain networks.* Changes in the *default mode network* (DMN) could explain co-morbid psychiatric symptoms and might predict treatment outcomes, while an altered *inhibitory control network* (ICN) may be why Internet addicts have trouble stopping and controlling overuse. Basically, this study supports earlier research that speaks to brain mechanisms that are involved in maintaining excessive and dysfunctional use, as well as why Internet addicts have trouble stopping once started.

Tech Stress Syndrome: Elevated cortisol levels

Years ago, I coined the term *Tech Stress Syndrome* to describe a physiological and psychological set of symptoms that resulted from heavy and addictive screen use. Obviously, the mind and body are not separate, and what you do affects your entire being. Behavior that affects your physiology is bound to affect your psychological well-being, and vice versa. Tech Stress Syndrome is most likely related to elevated cortisol levels for an extended period. We are not designed to experience protracted periods of arousal, attention, and focus that are often associated with excessive screen and Internet use.

Technology is also big business. I often find myself more stressed around the near-constant upgrading and updating of hardware, devices, software, and apps. It feels like a never-ending merry-go-round where you can never find yourself on firm ground. It's stressful always feeling you must be up to date and having to make sure you have the latest and greatest device or operating system — I've never found it to appreciably change what I do, nor does it seem to make any notable changes in my quality of life. This is the essence of Tech Stress Syndrome, where there is an industry-supported focus on *new* and *improved* versions of everything!

WARNING

Extended online periods of attention and concentration, with *always-on readiness*, are not normal for human beings. People are designed to be able to respond (with elevated psychological and physiological arousal) for moments or short periods of readiness to attend and act, but not hours or days on end. Think of our smartphones, where we are "on" all the time and never "off." This constant stress leads to an unending arousal and adrenal activation and a resulting elevation of cortisol, which is a function of a *readiness to respond* to things on our phones and devices, as well as those viewed on the Internet. This hypervigilance can be a response to a video game, an email, a text, a social media update, or scrolling/surfing, and all the communication modalities that stream or are pushed through to us every day. The issue is not *one thing*, but rather an unending *stream* of things. There is no end to the pipeline of stimulation and to the attention this stimulation requires to process. This is part of Tech Stress Syndrome, and I am afraid many people are experiencing it without even knowing it.

People need a break from the endless stimulation of their screens. Think about it. Is there ever a time you are without one of them? If you are not at work or school with a computer or laptop, you are streaming TV, and all the while you still have your smartphone in your pocket and are always dialed into the Internet. There seems to be no time that you are not onscreen or screen-ready. With an Internet-addicted person, this can be amped up to a point where there is literally no time off-line, and along with reduced sleep, the brain and body don't get time to really rest.

REMEMBER

Always-on readiness to respond equals stress.

Sleep and circadian rhythm disruption

Sleep problems may be the most prevalent issue with both normal and addictive overuse of screens. Why is sleep relevant when it comes to your screen use? Well, the most obvious reason is that the sheer amounts of time people spend on their screens can rob them of precious sleep time. There are limited hours in a day, and as I discuss in Chapter 4, you can lose track of time when you're online, but someone with an Internet addiction will become so involved online or in a video game that they will stay online until the early morning hours and then have problems waking up for school or work the next morning.

WARNING

You can't go without sufficient sleep for very long. People experience significant physical, psychological, and medical problems from too little or poor sleep. The negative effects are numerous and include hypertension, headaches, impaired attention and concentration, irritability, emotional volatility, increased aggression, depression, lowered immune response, colds and flu, weight gain, memory changes, impaired motor skills and coordination, and more. The effects of lack of sleep are cumulative over time, and you build a *sleep debt* that needs to be paid.

Many screens are kept in the bedroom, along with a smartphone, which is essentially a portable Internet connection; these can and do keep you awake for hours when you should be sleeping. You can access all the world's stimulating content from your bed, and although you may think it is relaxing, it is actually elevating cortisol (a stress hormone) and disrupting your circadian sleep rhythms. The blue light from screens, sounds, dopamine activation, and stimulating content are *all* very disruptive to sleep. Even having your phone off, but near you in bed, can be sleep-disrupting.

REMEMBER

Sleep deficits build over time and can create a sleep debt that must be paid back to your body. You will eventually need to repay yourself for a lack of sleep — your body remembers the sleep you missed.

It's a cycle: The circadian rhythm

Circadian rhythms are natural sleep and wake cycles that are also related to physical health. They are particularly responsible for alertness during daytime hours and drowsiness during evening and sleep-time hours. They are natural rhythm cycles that are linked to deep evolutionary development of the pineal gland in the brain. Several hormones are associated with sleep and with maintaining our circadian rhythms (as you can see in Figure 10-1), and heavy screen use seems to interfere with these natural cycles, especially during evening and sleep-time hours. Some theories state that blue light emitted by LCD screens has an impact on melatonin production, which is mediated through the eyes and the pineal gland. Screen light seems to disrupt this natural melatonin conversion cycle and disrupts normal biological processes.

TECHNICAL STUFF

It's interesting to note that melatonin is a metabolic precursor to serotonin production. As you may know, serotonin and other precursors such as L-tryptophan are very involved with mood and, to some extent, sleep. Serotonin is often at least partially implicated in depression, along with dopamine and norepinephrine. Other, less well-understood neurotransmitters are undoubtedly implicated in sleep, and are also likely involved in mood regulation.

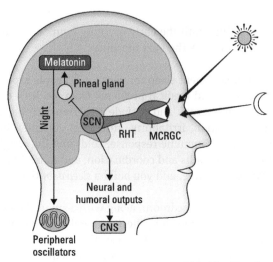

FIGURE 10-1:
The circadian
rhythm.

© John Wiley & Sons, Inc.

Yes, there is such a thing as good sleep hygiene

Good sleep hygiene is a lot like good dental hygiene; you may not always want to do it, but if you don't, you'll feel the effects later. Sleep hygiene encompasses the entire biobehavioral cycle of sleep preparation, habits and bedtime rituals, room lighting, temperature, level of in-bed activity, screen use prior to going to bed, watching TV in bed, stress management, and more.

REMEMBER

It is generally not recommended to have any screens in your bedroom — even a TV. You want to build good sleep hygiene habits around sustainable patterns that foster good sleep. The bedroom should ideally be used for only two things, and one of them is sleep. Perhaps some bedtime reading, which can induce drowsiness, is okay, but TV viewing, laptops, working in bed, and spending time on your smartphone (and even using your smartphone as an alarm clock) are potentially problematic. The idea is to create an association in your brain that connects the bedroom with bedtime, relaxation, sleepiness, and sleep. Screen associations may have the effect of disrupting your sleep.

Looking at Behavioral Issues When You're Addicted to the Internet

Numerous neuropsychological and behavioral impacts are associated with heavy screen use and Internet addiction, as you find out in this section. Some of these functions can also be affected by other addictions, illnesses, and psychiatric disorders, but many of the deficits are specific to screen-based addictions. Heavy screen use does in fact change us.

Some of the behavior changes are innocuous and temporary, while others seem to be more negative and impactful. Most of these changes appear to be transient in the sense that the brain's neuroplastic ability to change, adapt, and adjust will allow new functional connections to develop over time, assuming the disruptive screen use pattern is changed. If the pattern is unchanged, however, these deficits can impact productive living.

Shortened attention span

A shortened attention span has been well documented in some studies that suggest *attentional capacity* can be disrupted while using a screen; you know this if you have tried to get someone's attention when they were staring at their screen. These studies also point out longer-lasting effects that seem to be brain-based. Neurons proliferate connections in the brain in the same way that we train a muscle to get stronger; the more you use the muscle, the more you build up muscle fibers, while the less you use it, the more likely it is that this same tissue will atrophy from disuse. In the brain, *neurons that fire together, wire together.* The more you use neural pathways and networks, the stronger those associated pathways and interconnections become.

REMEMBER

So, what this essentially means is that you must *use it or lose it.* If you don't practice and exercise your attentional capacity, you'll likely lose some of this ability, perhaps not permanently, but to a noticeable degree.

All things screen-based, which of course include the Internet, video games, and smartphones, can impact attentional capacity. This impact is certainly noticeable while you're onscreen, but it can also create brain-based changes in how you think and process information. It is important to note that a user or addict's perception of their ability to attend is often distorted and inaccurate. People might think that they are attending, but studies show their ability to attend and comprehend is nevertheless negatively impacted.

REMEMBER

We often confuse attention and memory. However, the reality is that what some people attribute to memory is due to attentional factors. *If you do not attend to something, then you do not fully encode the memory for later recall or recognition.* There is also a difference between *intention* and *attention*. You can want to or need to attend to something important (intention), but the reality is that attention is a neurological process and cannot be faked. If you are not really attending, you will miss whatever is going on. Screens are amazing at fooling our intention and captivating our attention despite whatever else is going on around us.

Anything that interferes with attention, such as the Internet and screens, can indirectly impact memory. Obviously, a small degree of inattention is normal, but the Internet, video games, smartphones, and essentially all things screen-related are potent attention grabbers. They are designed to *pull* and *hold* your attention and, as such, are also addictive. In many cases they are so stimulating and addictive that a person is unaware that they are psychologically absent in conversations or other activities that require their attention.

REMEMBER

It is said that we now live in the *attention economy*, where the most valuable currency we have is our time and attention. No wonder the tech and media industries spend billions to capture that asset. No one gets more time, and unless we take personal control over it, we will lose this asset.

Impatience and instant gratification

I am what I call a *halfway digital native*. What I mean is that I spent half my life without Internet and screen technology; in fact, I got through my undergraduate, graduate, internship, and residency training without really using a personal computer, which was just becoming popular in medicine and hospitals. I remember being on residency and starting to use one in 1987 at the hospital where we still used paper charts. A lot has changed.

My children, on the other hand, have grown up with all the latest innovations, including the personal computer, fast Internet, laptops, Wi-Fi, and of course, the smartphone. My observations here are far from scientific, but numerous studies support this. Because of the near-instant access to all the world's information, as well as instant communication, there is less patience and more expectation of instant gratification. Time may mean something very different to millennials and Gen Z as well as the latest tech-savvy users. Time has been warped by the digital shortcuts that have been woven into everyday life. All information now has a time-based half-life, so the *immediate* or *instant* is seen as more valuable than something older.

If I try to explain something in detail to my son, he might say, "I know, Dad," or "I'll just look at YouTube." Now, if he would just listen for two minutes (I really do sound like a boomer here), I could explain what he wanted to know, but the reality is that he, like many others of his generation, does not want to take the time; seconds are like hours to them. More specifically, they do not take the time to learn in the *analog format* that I and all predigital natives valued. Some might argue that this is a reasonable alternative way of human learning and cognitive functioning, although one cannot help but wonder if something is lost when priority is given only to the *rapid result* and very little to the *process* of getting there. You can see my age showing.

Some research supports the idea that our brains have changed, that we attend less well and have less patience, and that we want (need) everything instantly or we move on. Computers, smartphones, and the ever-on, powerful Internet have changed us both functionally (behaviorally) and neurologically (structurally). If this were a different publication, I might speak to the fact that humans are evolving to a more human-machine interactive species, where we are in essence becoming "cyber" — with the use of these devices, all connected to the Internet, contributing to a new, augmented human experience.

WARNING

The problem is that patience and the ability to delay gratification are very necessary life skills. The ability to delay gratification is essential in being able to accomplish any long-term goal. Even if we are migrating to a world where we relate to our devices as an augmented part of our cognitive functioning, we are still human, and addiction to this technology can and does interfere with life balance (hampering our ability to accomplish our longer-term goals).

Distraction issues (especially while driving)

Distraction is the result of inattention. What you do not fully attend to has the leftover effect of producing distraction. This is often an annoyance to family, co-workers, and loved ones; at other times, it can prevent you from completing duties at school or work. However, there is one place where inattention, and resulting distraction, can be deadly — in the car.

Driven to distraction

Many tasks require some degree of attention. Some of these tasks are innocuous, in that there will be no physical danger if you're inattentive or distracted by a screen. However, one exception might be when operating an automobile.

Driving a car requires a lot of human computing power. The problem occurs when you engage in any activity that is highly distracting while driving, such as operating a smartphone. A considerable amount of research has been conducted into the distractive ability of smartphones and how much cognitive and attentional bandwidth they consume while a person is driving.

Driving a car is a highly complex neuropsychological process that involves monitoring multiple information streams and constantly executing real-time course and error corrections. Because the vehicle is moving rapidly, conditions change quickly, and reaction time, visual-spatial processing, fine and gross motor skills, memory, and cognitive processing must all be readily accessible.

THE PHYSICS OF DISTRACTED DRIVING

You travel 97.5 feet per second at 65 mph, with a stopping distance of approximately **125 to 150 feet** at that speed.

The **minimum average time** to experience the impulse to pick up your phone or respond to a notification trigger and neurologically process it and respond is approximately **3 to 6 seconds or longer.**

You can therefore inattentively **travel 300 to 600 feet** during that time, thereby increasing accident risk by **600 to 1,000 percent (that is a six to ten times greater accident risk).**

You are about **23 times** more likely to have an accident if you operate a smartphone while driving, especially texting or using other apps. You are at a **six times** greater risk of having an accident from distracted smartphone driving than from alcohol intoxication. I call this *intextication*.

Due to distracted smartphone driving, **1,600,000** accidents happen each year (according to the National Safety Council). There are **330,000** injuries each year due to smartphone distraction (according to a Harvard Center for Risk Analysis study), and **7 percent** of all fatal crashes in 2019 were distraction-related, resulting in **3,142** lives lost, an increase of **9.9 percent** over 2018, when **2,858** lives were lost due to distracted driving, and nearly **25 percent** of *all* car accidents involve smartphones in some way, according to the National Highway Traffic Safety Administration (NHTSA).

Lastly, driver distraction is responsible for more than **58 percent** of teen crashes, according to the AAA Foundation for Traffic Safety. Teen and young adult brains are even more vulnerable to smartphone-related distraction when compared to adults.

WARNING

Your ability to accurately appraise your relative attentional impairment is distorted when you're using your phone while driving. This is very similar to the findings based on research conducted on drivers' subjective judgment of impairment while they are alcohol intoxicated. In most cases, the more intoxicated the person was, the more impaired they were, and the less capable they were of recognizing those limitations. This distortion is exactly what we see with smartphone use while driving; you may not recognize your own attentional impairment and distort your relative ability while driving and using your smartphone.

I've consulted on research and public safety efforts on this issue, and it is perhaps the one definitive area of Internet addiction that has frequent fatalities connected to it. It's impossible to console someone who has lost a loved one because a driver used their smartphone as they crossed the median into oncoming traffic.

The driving force

We all know that smartphones can be addictive, and that we can and do overuse them to a point of addiction that results in harm to parts of our lives. Here we have a competition of sorts. It's an attentional competition, where your use of the phone while driving pulls necessary attention away from operating the vehicle. The real issue here is that if the smartphone is addictive and compelling while stationary, why would it stop being addictive while you are driving? How many of us have picked up our phones while driving (even though it is illegal in most states to do so)? I know I have.

The more addicted you are to your phone, the more likely you will use it while driving. Without being dramatic, it is like Russian roulette, in that at some point, there is a statistical likelihood that something will happen. You can roll the dice many times without hitting a seven, but eventually a seven will emerge. It is the illusion of control over yourself and your vehicle that contributes to the cognitive distortion that you are safe, despite not being so.

I did a study called *It Can Wait* a few years back (2014 and 2015) in collaboration with AT&T, and some of the data I analyzed showed that most people admit to knowing that operating a smartphone while driving is dangerous, but 75 percent of them admit to doing it anyway. The reason is twofold. First, smartphones are addictive, plain and simple, and second, there is little correlation between people's attitudes, beliefs, and ultimate behavior.

I hypothesized in the first study we did that people texted while driving (which is what the initial laws addressed), but I also knew it was likely that people were doing a lot more than that. In the follow-up study, I found that 62 percent kept their smartphones within easy reach while driving. Thirty percent of people admitted to posting to Twitter while driving. Twenty-two percent who accessed social networks while driving cited addiction as a reason. Of those who shot videos behind the wheel, 27 percent thought they could do it safely while driving! People admitted to doing essentially all the land-based phone activities they would normally engage in while driving, including checking Facebook, taking photos, making videos, scrolling, and banking.

About 30 percent of the 1,000 subjects (ages 16–65) in our study felt that using their smartphone while driving was a habit, while about 11 percent felt they were addicted to their smartphones. Also notable was that about 20 percent of study participants felt compelled to respond to a text when it came in while they were driving. This supports the trigger hypothesis I mention earlier in this book, where a notification elicits a significant release of dopamine that kindles the compulsive checking response followed by another, lesser hit of dopamine. It also supports research that showed that notifications elevate cortisol (the stress hormone) and that a way to lower that discomforting hormone level is (you guessed it) to check your phone.

Multitasking myths

When I do trainings and lectures, I ask the audience whether they believe in multitasking; some do and some don't. Sometimes I playfully ask how many women in the audience think that they can do it, but that men cannot. This usually gets a laugh, but it makes the beginning of a point.

REMEMBER

Multitasking *does not* really exist. You cannot, I repeat *cannot,* process two attentional streams at once. You can switch from one thing to another, sometimes rapidly, but you can attend to only one thing at a time.

I remember my kids doing homework when they were younger. They sat on the floor with a couple of books, a notebook, a smartphone, a laptop, the TV on, and some food and a drink. I would naively ask what they were doing, and they would say their homework (oh, and they had headphones on, too). They may have been doing their homework, but they were doing a whole lot of other things, too. They probably did eventually get their homework done, but I know it took them a lot longer. I also seem to remember being told that "our generation can multitask, unlike you boomers." While there may be differences between millennials and boomers, I am sure that multitasking is not one of them. It simply takes longer to reach the same level of single-task focus and comprehension when attempting to multitask.

If you try to process multiple things at once, there is about a 40 percent reduction in comprehension efficiency. Check out the book *Human Capacity in the Attention Economy* (published the American Psychological Association); Sean Lane and Paul Atchley do a great job as editors looking at all the issues with attention (I also have a chapter in that book). Switching from task to task can lead to errors and amplified distraction, as well as reduce your task accuracy. It looks like we can do things simultaneously, but this is an illusion; to test this, next time you have two or three tasks on your desk, try to do more than one at a time.

REMEMBER

Our screens are powerful attention grabbers as well as a big distraction. They are ever-present, and we assume we must use them for everything when often we need to put them aside (perhaps even turning them off), especially when we want to get something done. If you must use the Internet or a screen for work or school, try to limit accessing other content to minimize distraction and attempts to multitask.

Decreased academic and work productivity

The Internet and computer screens have a positive aura around them, and we associate productivity with their use. We assume that all this technology makes us more efficient, smarter, more easily educated, and more productive in school and the workplace. The problem is that this is simply not true.

Numerous studies have looked at the use (and overuse) of the Internet and video games and their effect on academic performance. A 2021 Rutgers University study, which was reviewed in *Science Daily*, found that test scores, motivation (remember reward deficiency syndrome?), and educational aspirations drop if children overuse noneducational screen technologies. This is not surprising, and the study's recommendation was no more than one hour daily and four hours on weekend days of nonacademic screen time. These numbers are even lower than the American Academy of Pediatrics two-hours-a-day recommendation. The Rutgers study also found that children who spent less than one hour a day of nonacademic Internet screen use had less boredom at school. This makes sense and correlates well with my research and clinical experience, where children, teens, and young adults who are heavy screen users become intolerant of boredom.

REMEMBER

Computers, screens, and the Internet look like they make us more productive; they are amazing tools so in some ways they do. The problem is that they are also *addictive and distracting.* They can easily be toggled between our work or school tasks and our fun tasks. It is estimated that upwards of 80 percent of the time spent on screens is not for productive purposes; even if schools and workplaces use blocks, monitors, and filters, people now have always-on computer screens hooked to the Internet in their pockets — their smartphones.

Even if all these screens and tech made us work more efficiently than before, if you subtracted lost productivity from downtime due to tech support, malfunctioning hardware, glitched software, Internet slowdowns, server outages, forgotten passwords, updating, and most of all, nonproductive distracted use of tech while at school or work, it might look different. I would say it may be a wash, if not hands down a net loss (no pun intended), but we still love our tech, despite having to enter in our failed password attempt for the tenth time that week and after spending three hours on hold with tech support. The Internet *is* a great thing, but great things won't stay great if you let them *run* your life.

About 20 years ago, I did a corporate study for Websense looking at web filtering, blocking, and monitoring software for the Internet. It's hard to believe that there was a point that the Internet was open and free, with no blocked websites in most workplaces. This was before the advent of modern smartphones, but even then, our research found that about one out of five days of the week was wasted on non-productive use of the Internet. Frankly, I think this number was low. This research was part of the initial movement to educate corporate America that the Internet was powerful and addictive, and that left to our own devices, we would never leave our devices.

Chapter **11**

Taking a Self-Assessment

O ver the years, I've developed and fine-tuned several screening instruments to help assess problems with Internet, video game, and smartphone use. Times have changed, so many of the tests in this chapter have been modified and updated. Some tests were initially based on my original published research in the late 1990s, which looked at Internet addiction and paralleled the adapted criteria for pathological (compulsive) gambling from the *Diagnostic and Statistical Manual of Mental Disorders*, 5th Edition (DSM-5). Since those early days, I've modified the instruments with expanded criteria, but the main point to remember is that addiction or overuse is generally a problem when it impacts real-time living in some notable way.

For scoring purposes, I suggest a conservative cutoff for recognition of a potential problem. Most of the instruments used for assessing and diagnosing behavioral addiction have a cutoff at approximately 35 percent; in other words, if you answer in the positive directions for admitting to a problem 35 percent of the time, then you would be meeting the problem cutoff criteria. For the instruments in this chapter, I've set the cutoff at approximately 50 percent; I'm using this higher percentage to allow for false positives and to address the fact that many of us overuse our Internet-based screens at times. The other thing to keep in mind is that these tests are looking at a particular point in time or "state" — they do not address long-standing or complex patterns or "traits," although repeating the test later will provide some increased validity in measuring behavioral patterns.

WARNING

Typically, the problem of Internet and screen addiction shows up in the form of impaired school or work performance, increased social isolation from family and friends, personality changes, irritability, sleep pattern disruption, reduced physical activity, and poor self-care. This also includes anger when access is interrupted to a video game, the Internet, or other screen devices. Perhaps the most important behavioral change is that a person takes little or no pleasure in any other activities except their screens and devices.

REMEMBER

These tools help you assess yourself or your loved one, but they are *not* hard-and-fast diagnostic tests. Any attempts at medical and psychiatric diagnoses must be performed in the context of a professional evaluation or consultation by a psychologist, psychiatrist, or other licensed mental health or addictions professional — preferably one who has experience in the diagnosis and treatment of Internet and screen-based addictions. These screening tools are for educational and informational purposes only, and *do not* constitute a medical opinion. Be careful not to take a test (or give a test to a loved one) and assume there is a problem without confirmation from other information and direct professional opinions. Please do not use these tests to prove to your loved one that they, in fact, do have a problem — in my experience, this simply increases the addict's denial and defensiveness.

I'll include a word here about professional help for assessment and diagnosis. Internet and technology addiction is a relatively new area of research and clinical focus in addiction medicine — it's barely 25 years old. As a result, not all doctors and therapists have knowledge and adequate expertise in the subject. Some disagreement still exists as to the etiology, epidemiology, diagnosis, and treatment of these issues, and although there is a consensus that Internet addiction is a problem in need of treatment, differences still exist as to how to diagnose, label, and treat Internet, video game, and other screen addictions.

We have come a long way, though. When I started in this area of study and clinical practice in the late 1990s, there was almost no published, peer-reviewed research or books on the subject. Now, there are perhaps a thousand or more. I suspect that as digital screen technologies change, so will our understanding of their management and treatment. One thing is for certain: The Internet and screens are not going away, and if a human being is involved in the process, they will need to address potential addiction and balanced living.

REMEMBER

Although I don't have a formal test to address an addiction to binge watching and online entertainment (see Chapter 6), their features aren't dissimilar from other forms of screen use. Typically, when I see binge-watching issues, they are in addition to other Internet and screen-based problems. I most often see them as a treatment concern when other screen access has been cut off or significantly reduced, and people gravitate to the most accessible form of screen activity. This can be an issue with Internet-based streaming sites such as YouTube, Amazon Prime, Hulu, and Netflix.

Warning Signs of Internet Addiction in Your Spouse, Friend, or Loved One

Recognizing the general warning signs of Internet addiction in your spouse, friend, or loved one is the first step in helping them help themselves. The following warning signs should serve as general guidelines for you to determine whether your spouse, child, family member, or friend may have a problem.

Do they

1. Spend a lot of time alone with their computer, tablet, or smartphone on a regular basis?

2. Become defensive, evasive, or angry when you confront them about their behavior?

3. Either seem unaware of what they have been doing or attempt to deny it?

4. Prefer spending time with their device, smartphone, or video game rather than with other people?

5. Lose interest in other previously important or enjoyable activities, such as friends, sports, work, school, hobbies, or exercise?

6. Appear to be more socially isolated, moody, or irritable?

7. Seem to be establishing a "second life," with new friends or people they met online?

8. Spend greater amounts of time online and attempt to cover or minimize the screen or hide the phone when you come into the room?

9. Talk about their activity on the Internet a lot, or avoid discussing it and seem secretive about their screen activities?

10. Exhibit signs that their work or school performance is suffering (perhaps they were fired, their grades or extracurricular activities are slipping, they are reducing their self-care, or their household responsibilities are neglected)?

11. Have legal problems because of their Internet porn behavior: the loss of child custody, divorce, arrest, or issues at work or school due to viewing or downloading pornography?

12. Appear to be oblivious to any negative impact that their screen use is having on their life?

If you answer positively to six or more questions, then consideration of a possible problem can be made.

Virtual Addiction Test

I coined the term *virtual addiction* in 1998 to describe the then-new phenomenon of Internet addiction, and it was also the title of my first book published the following year. There is likely to be some overlap across many, if not most, of the tests in this chapter, although there are some subtle differences in the wording and focus of what is being assessed. The Virtual Addiction Test is general and meant to address overall use of the Internet and screens; it isn't focused on a particular content or device. This test may be used for general screening (no pun intended) purposes. See Chapter 4 for an introduction to the addictiveness of the Internet and technology.

1. Do you lose track of time when using the Internet or your smartphone?

   - Yes

   - No

2. When you're not on the Internet or your smartphone, are you still thinking about using them?

   - Yes

   - No

3. Do you find that you are spending a greater amount of time on the Internet or your smartphone?

   - Yes

   - No

4. Do you find yourself repeatedly seeking more stimulating or newer content on the Internet or your smartphone?

   - Yes

   - No

5. Have you made repeated but unsuccessful efforts to control, limit, or cut back your Internet or smartphone use?

   - Yes

   - No

6. Do you find yourself restless, anxious, or irritable when you do not have access to the Internet or your smartphone?

   - Yes

   - No

7. Do you feel you are using the Internet or your smartphone as a way of escaping or relieving a negative mood (characterized by, for example, boredom, frustration, stress, anxiety, or depression)?

- Yes
- No

8. After spending an excessive amount of time on the Internet or your smartphone, and then trying not to do so the next day, do you find yourself using it again right away?

- Yes
- No

9. Do you avoid sharing the extent of your involvement with the Internet or your smartphone with friends, family members, or others?

- Yes
- No

10. Have you ever committed illegal acts related to your use of the Internet or your smartphone?

- Yes
- No

11. Has your use of the Internet or your smartphone ever jeopardized or impacted a relationship, job, or educational opportunity?

- Yes
- No

If you answer positively to five or more questions, then consideration of a possible problem can be made.

Smartphone Compulsion Test

This test focuses more exclusively on the smartphone and smartphone use. While the smartphone would likely not be an issue were it not for it being connected to the Internet, it has been a game changer in terms of increased access, availability, and overuse of the Internet. Chapter 4 has details on what makes smartphones so addictive.

1. Do you often find yourself spending more time on your smartphone than you realize?

   - Yes

   - No

2. Do you often find yourself mindlessly passing time by endlessly scrolling on your smartphone?

   - Yes

   - No

3. Do you seem to lose track of time when you are on your smartphone?

   - Yes

   - No

4. Do you find yourself spending more time on Snapchat, Instagram, Twitter, Facebook, and TikTok — texting, tweeting, posting, commenting, instant or direct messaging, emailing, or scrolling through social media — as opposed to interacting with people in real life?

   - Yes

   - No

5. Has the amount of time you spend on your smartphone been increasing?

   - Yes

   - No

6. Do you wish you could be less involved with your smartphone (for example, not picking it up or looking at it repeatedly)?

   - Yes

   - No

7. Do you sleep regularly with your smartphone (turned on) under your pillow or next to your bed?

   - Yes

   - No

8. Do you find yourself viewing and answering texts, tweets, emails, instant messages (IMs), direct messages (DMs), Instagram, TikTok, Snapchat, or social media at all hours of the day and night — even when it means interrupting other things you are doing?

- Yes

- No

9. Do you text, email, tweet, instant message (IM), direct message (DM), Snapchat, TikTok, YouTube, or surf while driving or doing other similar activities that require your focused attention and concentration?

- Yes

- No

10. Do you feel that your smartphone use decreases your productivity?

- Yes

- No

11. Do you feel reluctant to be without your smartphone, even for a short time?

- Yes

- No

12. Do you feel uncomfortable when you accidentally leave your smartphone in the car or at home, have no service, or have a broken phone?

- Yes

- No

13. When you eat meals, is your smartphone often on the table or easily accessible?

- Yes

- No

14. When your smartphone rings, beeps, or vibrates, do you feel an urge to check it and look at it?

- Yes

- No

15. Do you find yourself mindlessly checking your smartphone many times a day, even when you know there is likely nothing new or important to see?

- Yes

- No

If you answer positively to seven or more questions, then consideration of a possible problem can be made.

Digital Distraction Self-Test

The Digital Distraction Self-Test is focused more on general themes of attention, time distortion, and lost focus related to the Internet and screen use. It is also designed to be a self-test as opposed to being completed by another person. Check out Chapter 4 for more about the addictiveness of the Internet and smartphones.

1. Do you spend more time online or on your digital screen devices (computer, laptop, tablet, or smartphone) than you realize?

- Yes
- No

2. Do you mindlessly pass time on a regular basis by staring at your smartphone, tablet, or computer, even when you might have better or more productive things to do?

- Yes
- No

3. Do you seem to lose track of time when you're on any of these devices?

- Yes
- No

4. Are you spending more time with virtual or social media "friends" as opposed to real people?

- Yes
- No

5. Has the amount of time you spend on your smartphone and the Internet been increasing?

- Yes
- No

6. Do you secretly wish you could be a little less wired or connected to your devices?

- Yes
- No

7. Do you regularly sleep with your smartphone (turned on) under your pillow or next to your bed?

- Yes

- No

8. Do you find yourself viewing and answering texts, tweets, IMs, DMs, Snapchat, TikTok, Instagram, Facebook, and emails at all hours of the day and night — even when it means interrupting other things you are doing?

- Yes

- No

9. Do you text, email, tweet, IM, DM, Snapchat, Facebook, Instagram, TikTok, or surf/scroll while driving or doing other similar activities that require your focused attention and concentration?

- Yes

- No

10. Do you feel that your use of technology decreases your productivity at times?

- Yes

- No

11. Do you feel uncomfortable when you accidentally leave your phone or other Internet screen device in the car or at home, or if you have no service or your device is broken?

- Yes

- No

12. Do you feel reluctant to be without your smartphone or other digital devices, even for a short time?

- Yes

- No

13. When you leave the house, do you always have your smartphone or other digital screen device with you?

- Yes

- No

14. When you eat a meal, is your smartphone always part of the table setting?

- Yes

- No

If you answer positively to seven or more questions, then consideration of a possible problem can be made.

Greenfield Video Game Addiction Test

The Greenfield Video Game Addiction Test (GVGAT) was developed with video game use and addiction in mind. Many, if not most, of the patients I have consulted with or treated have had some issue with video game use, so a test focusing specifically on this issue was needed. See Chapter 7 for more about video games.

1. Do you feel a loss of control (inability to stop or limit playing) when using video games?

- Yes

- No

2. When you are not using video games, are you preoccupied with the games or gaming (for example, thinking about them, reliving past experiences, planning the next time you will play, or thinking about when or where you will next have your next access to the games)?

- Yes

- No

3. Do you find that you need to spend greater amounts of time on video games to achieve the same satisfaction?

- Yes

- No

4. Do you find yourself seeking more stimulating (that is, exciting, new, or more challenging) video games?

- Yes

- No

5. Have you made repeated but unsuccessful efforts to control, limit, or cut back your video game use?

- Yes

- No

6. Do you find yourself restless or irritable when attempting to (or forced to by a parent) cut down or stop using video games?

- Yes

- No

7. Are you using a video game as a way of escaping from problems or relieving a negative mood (characterized by, for example, boredom, frustration, fear, anxiety, anger, or depression)?

- Yes

- No

8. After spending what you consider an excessive amount of time on a video game, and vowing not to do so again, do you find yourself playing again soon after?

- Yes

- No

9. Do you find yourself lying to family members, friends, doctors, therapists, or others to conceal the extent of your involvement with video games?

- Yes

- No

10. Do you find yourself committing illegal, self-harmful, or self-defeating acts related to your use of video games?

- Yes

- No

11. Have you jeopardized, impacted, or lost an academic or educational opportunity, relationship, or job because of your use of video games?

- Yes

- No

12. Has your work or academic performance been impaired as a direct or indirect result of playing video games?

- Yes

- No

13. Have you experienced any health or medical problems because of your video game use?

- Yes

- No

14. Do you watch YouTube or other video streams or recordings of people playing video games?

- Yes

- No

15. If your parent or someone else removes your computer, tablet, smartphone, or gaming console, do you experience anger, sadness, or irritation?

- Yes

- No

16. Have you ever experienced a loss of sleep related to your video game use?

- Yes

- No

17. Do any of your friends or family think you have a problem with your video game use?

- Yes

- No

18. Do you lose track of time when playing video games?

- Yes

- No

If you answer positively to eight or more questions, then consideration of a possible problem can be made.

Shopping, Gambling, and Investing

No formal tests exist for looking at how online shopping, gambling, and investing (covered in Chapter 8) can affect people. In most cases, the guidelines for real-time gambling are relevant and applicable for online gambling. Some of the issues that arise from these three activities may be addressed by the 12 warning signs described earlier in this chapter.

The following list presents some of the well-established diagnostic criteria from the DSM-5 for pathological gambling. These criteria apply nicely for looking at online gambling, stock trading, and shopping, although they are not a perfect fit. I would use these criteria along with the 12 warning signs assessment (found earlier in this chapter) to addresses these areas of concern.

Persistent and recurrent problematic gambling behavior leads to clinically significant impairment or distress, as indicated by the person exhibiting four (or more) of the following types of behavior in a 12-month period:

a. Needs to gamble with increasing amounts of money to achieve the desired excitement

b. Is restless or irritable when attempting to cut down or stop gambling

c. Has made repeated but unsuccessful efforts to control, cut back, or stop gambling

d. Is often preoccupied with gambling (for example, having persistent thoughts of reliving past gambling experiences, handicapping, or planning the next venture, or thinking of ways to get money with which to gamble)

e. Often gambles when feeling distressed (for example, when feeling helpless, guilty, anxious, or depressed)

f. After losing money gambling, often returns another day to get even ("chasing" one's losses)

g. Lies to conceal the extent of involvement with gambling

h. Has jeopardized or lost a significant relationship, job, or educational or career opportunity because of gambling

i. Relies on others to provide money to relieve desperate financial situations caused by gambling

Online shopping and stock trading are also relatively new areas of research and study. I suspect these will be areas of increased focus in the coming years because so much of our shopping, investing, and commerce has moved online. Stock trading and auction sites (like eBay and other sites catering to auctions) have a lot in common with online gambling. The idea that you have endless opportunity and the ability to impulsively click to buy can make online behavior with these areas more problematic and more addictive.

WARNING

The problem with both shopping and stock trading is that they are hidden beneath the veil of necessity and practicality. I know I experience a sense of comfort and ease when it comes to shopping and trading online, and there have been times that my Amazon or auction behavior was problematic. It's just too easy to pick and click without the direct and immediate consequence of what's being done. I haven't yet seen many patients who have a primary issue with online investing, but I have seen several whose online shopping has reached compulsive proportions. Obviously, if you or a loved one seem to have an issue with one of these areas, then getting a consultation with a qualified professional might be warranted.

Online Pornography Test

The Internet is a near-perfect medium for the production, distribution, and access of pornographic content. The ease of access and availability make online pornography addiction one of the top reasons people seek help for their Internet addiction behavior. Chapter 9 has more information about online porn addiction.

1. Are you spending more time viewing online pornography on your screen devices than you realize (computer, laptop, tablet, or smartphone)?

- Yes

- No

2. Do you mindlessly pass time on a regular basis on your smartphone, tablet, or computer by surfing, scrolling, viewing, or downloading pornography?

- Yes

- No

3. Do you spend more time viewing or downloading pornography as compared to engaging in real-time social or sexual activities?

- Yes

- No

4. Has the amount of time you spend watching pornography on the Internet been increasing?

- Yes

- No

5. Do you wish you could be less involved with online pornography?

- Yes

- No

6. Do you sleep with your smartphone, tablet, or laptop near your bed or under your pillow or next to your bed to view pornography?

- Yes

- No

7. Do you view pornography at all hours of the day and night — even when it means interrupting other things you are doing, including sleep, schoolwork, or work?

- Yes

- No

8. Do you look at pornography while you are driving?

- Yes

- No

9. Does your use of Internet pornography sometimes decrease your productivity at work or school?

- Yes

- No

10. Do you feel uncomfortable when you accidentally leave your smartphone or Internet device in the car or at home, when you have no service, or if your screen device is broken?

- Yes

- No

11. Do you feel reluctant to be without Internet pornography, even for a short time?

- Yes

- No

12. Do you typically masturbate while (or after) you view online pornography?

- Yes

- No

13. Do you vow to stop or reduce using online pornography and end up using it again soon after?

- Yes

- No

14. Have you ever looked at pornography sites that make you uncomfortable?

- Yes

- No

15. Have you had legal problems because of your Internet pornography behavior — for example, the loss of child custody, divorce, arrest, or work issues due to viewing, downloading, or sharing pornography?

- Yes

- No

16. Do you find that you have less sexual arousal or interest in sex with a partner when you regularly use pornography?

- Yes

- No

17. Have you ever experienced any difficulty in achieving an erection or becoming sexually aroused because you watched too much pornography?

- Yes

- No

18. Have you ever used online pornography at school or at work?

- Yes

- No

If you answer positively to eight or more questions, then consideration of a possible problem can be made.

Cybersex Abuse Test

Cybersex means a more general use of the Internet and screens for sexual activities and behaviors that include pornography, webcams, OnlyFans, sex hook-up and dating sites, websites and apps for sex work (escorts and prostitution), online strip clubs, phone and video sex, arranging for real-time or online affairs, and so on. This test covers most forms of online sexual behavior or the use of the Internet for arranging or facilitating real-time sexual contact. See Chapter 9 for more information.

1. Do you spend a large amount of time in online sexual or pornographic chat rooms, webcam sites, OnlyFans, phone sex sites, or sexual meet-up or affair sites, or engage in subtle or explicit online sexual conversations with or without masturbation?

- Yes

- No

2. Do you gravitate toward one or more individuals with whom you have regularly scheduled or frequent unscheduled online sexual contacts?

- Yes

- No

3. Do you become more depressed, isolated, or lonely as you spend more time pursuing online sexual activities?

   - Yes

   - No

4. Have you made attempts to have face-to-face contact with individuals you met on the Internet (by video, phone, or physical meetings), and do these meetings or phone or video conversations involve overt sexual discussions, masturbation, or actual physical sexual contact?

   - Yes

   - No

5. Do you hide information from your spouse, significant other, friends, or family regarding your sexual activities on the Internet, and do you find yourself being secretive about the nature and extent of your use of the Internet for sexual activity?

   - Yes

   - No

6. Did you initially accidentally find sexual online situations, apps, or websites on the Internet (such as pornography, cybersex, chat rooms, personals, webcam sites, OnlyFans, or escort/prostitution sites), but now you find yourself regularly seeking these sites, apps, and activities?

   - Yes

   - No

7. Do you find yourself having frequent thoughts about pornography and other online sexual activities, behaviors, or real-time sexual contacts?

   - Yes

   - No

8. Do you find the anonymity, intense intimacy, loss of inhibition, and timelessness while having online sexual interactions to be more stimulating and satisfying than real-time sexual relationships?

   - Yes

   - No

9. Do you find it difficult to stop going online for sexual purposes, and do you feel compelled to do so frequently?

   - Yes

   - No

10. Have you made commitments or promises to others or yourself not to access sexual or pornographic websites, but do so anyway?

- Yes

- No

11. Do you experience guilt or shame about your sexual use of the Internet?

- Yes

- No

12. Do you engage in masturbation while online, at times to the exclusion or avoidance of sex with your partner or spouse?

- Yes

- No

13. Do you find that significant individuals in your life, including your partner or spouse, friends, or family, are becoming concerned with the amount of time and energy you are devoting to your sexual use of the Internet?

- Yes

- No

If you answer positively to six or more questions, then consideration of a possible problem can be made.

Child Technology Test: Is Your Child Too Connected?

This test is focused on parents or caregivers and observations of their children's Internet and screen behavior. Many times children and adolescents spend excessive amount of time of the Internet, often using their smartphone as their dominant Internet screen device (although computers, laptops, and tablets are also used). Smartphone Internet behavior has increased steadily over the last 13 years or so since the introduction of the smartphone. Since the smartphone is essentially a small, portable Internet screen device, it lends itself perfectly to child, tween, and teen behavior, although this behavior has also increased for younger and older adults as well. Chapter 15 has tips for parenting in the screen age.

1. Do you find your child spending more time online or on their screens (smartphone, computer, laptop, tablet) than they seem to realize?

   - Yes

   - No

2. Do you find your child mindlessly passing time on a regular basis by using their computer, laptop, tablet, video game, or smartphone — even when they have required or productive things to do?

   - Yes

   - No

3. Do you find your child spending more time with virtual or social media "friends" as opposed to real people?

   - Yes

   - No

4. Has the amount of time your child spends on screen devices and the Internet been increasing?

   - Yes

   - No

5. Do you wish your child could be a little less wired or connected online to digital devices such as smartphones, computers, laptops, tablets, and video games?

   - Yes

   - No

6. Does your child regularly sleep with their smartphone (turned on) under their pillow or next to their bed?

   - Yes

   - No

7. Do you find your child viewing and answering texts, tweets, IMs, DMs, Snapchat, Instagram, Facebook, TikTok, or other social media posts, updates, or comments at all hours of the day and night — even when it interrupts other things they are doing, such as schoolwork, eating meals, playing sports, sleeping, or engaging in other family activities?

   - Yes

   - No

8. Do you limit, block, control, or filter Internet access and screen-time access (on any devices) for your child?

- Yes

- No

9. Do you feel your child's use of technology reduces their academic productivity and/or real-time socialization, family participation, self-esteem, or physical activity?

- Yes

- No

10. Do you find your child feeling upset or uncomfortable when they accidentally leave their phone or other digital screen device in the car or at home, they have no service, their device is broken, or their device is taken away from them?

- Yes

- No

11. Do you feel your child is reluctant to be without their smartphone or other screen devices, even for a short time?

- Yes

- No

12. Do you find yourself feeling reluctant to limit or remove your child's use of smartphones, other Internet screens and devices, or video game devices?

- Yes

- No

13. When you limit screen access or take it away from your child, do they have a strong emotional or physical reaction?

- Yes

- No

14. Does your child seem to lose track of time when they are using a computer, laptop, tablet, video game, or smartphone?

- Yes

- No

15. Does your child play video games to the point that it seems to be impacting their life?

- Yes

- No

16. Does your child view YouTube or other video streaming sites to the point that it impacts their life?

- Yes

- No

17. Does your child regularly view online pornography?

- Yes

- No

If you answer positively to eight or more questions, then consideration of a possible problem can be made.

IN THIS CHAPTER

» **Knowing that life is too big to fit on a screen**

» **Undergoing a digital detox**

» **Watching your online time and content use**

» **Matching your values to your tech time**

» **Saying goodbye to addictive notifications and apps**

» **Using self-help worksheets, resources, and treatment options**

Chapter **12**

Adopting Self-Help Strategies

L et's face it: Addictions are human problems, and as I discuss in Chapter 2, they are a biological part of the human condition. Many people face an addiction, compulsion, habit, or abuse of a substance or behavior at some time during their lives. Although drugs and alcohol are the most typical addictions you hear about, many other addictive behaviors create challenges and leave many people struggling. Gambling, sex, food, and exercise are just a few, and since all addictions share the same underlying neurobiology, it isn't at all surprising that Internet, video gaming, and smartphone technologies can be addictive.

Self-help is about finding information, techniques, resources, treatment, suggestions, and support to help in the process of managing an addiction. People typically think of self-help when it comes to addictions — in part because the mental health and addictions fields have not had a stellar track record in effectively treating addictions. Over the last century, many self-help groups and organizations have provided valuable information, support, and inspiration to addicts and their loves ones.

The hard part about self-help is that it is generally left up to the addict to seek out when and if they are ready. However, many Internet addiction patients do not initially see themselves as having a problem serious enough to seek support for it. As with all addictions, friends and family are often the first to notice and to be affected by those who are addicted; these friends and family can significantly benefit from support as well.

REMEMBER

I would add that self-help is also about anything you can do to access help, information, and support in *any* part of your life — not just limiting it to recovery from an addiction. Some of what I present in this chapter is about general use and overuse of screens, as well as an addiction to Internet technology, and there are many people who overuse their screens but who may not meet the criteria for an addiction. I see Internet and screen problems as existing on a continuum, with use, overuse, abuse, and finally progressing to an addiction. I recognize that these labels may be based on somewhat artificial distinctions, but they capture the essence of what I have observed over the last 25 years in this field.

Remembering That Life Isn't Lived on a Screen

I love technology. This might be surprising to you, but I've always had an interest in science and electronics. When I was a young child, I used to drag a bunch of electrical wires around the house, and that was the beginning of a lifelong passion in electronics and gadgets. I still remember the countless radios, tape recorders (remember reel-to-reel?), and other things that I took apart — often not being able to put them fully back together. I should also probably admit that I had some of the earliest personal electronic organizers and a car phone (which weighed 15 pounds) as soon as it was available.

I suppose it makes some sense, then, that I would somehow combine my lifelong interest in science and technology with my career in addiction medicine, psychiatry, and clinical psychology. The merger of these areas of interest has allowed me to help my patients, while at the same time reminding myself of some of the most important lessons regarding technology and balanced living.

Screen technologies are very compelling. They are fun, interesting, and rewarding to your brain. With a promise of better living through technology, there are many ways in which modern life encourages you to use the Internet and screens for everything. It's easy to get caught up in the distraction, boredom avoidance, perceived convenience, and ease of access that these digital devices provide.

But there is another side to the equation. All the fun, avoidance of boredom, and convenience comes at a price — not just the price of overuse and potential addiction, but the price of losing your humanity and valuable time. The following sections will help remind you of the benefits of reducing your screen time.

Recognizing that it's tough to limit tech use

It's hard to limit something that feels limitless. After all, screen tech is everywhere and it's only expanding. People have faster processors, screens on everything in their homes, 5G wireless smartphones, and abundant messages from corporate media on how *good* it is to be always connected to everything. But is it *good* for you? The Internet will soon be available from all corners of the planet, and that can be a good thing, provided there is a connection between the technology and human side of the equation.

There is a place that I love to visit in the country that sits in the middle of the woods on 11 acres. When I first started going there in the early 1990s, there was no Internet; eventually we had limited dial-up Internet, but before that, if we wanted to get our email, we had to drive into town to find someone who had an open router to hack their Wi-Fi. Finally, we got DSL, a very slow one, but it allowed us to do a few more things, and just recently, 25 years later, we finally got fiber-optic high-speed Internet. The truth is that I feel conflicted about having Internet access. There was something nice about being off the grid and about being away from the din of the noise that technology brings.

Part of what the Internet brings is the expectation (more of a constant, subtle insistence) that we should all be on our devices all the time — but when we are, we have no time off-line. How can we rejuvenate when we work, shop, scroll, surf, and are entertained no matter where we are or what we are doing? Even the news is interspersed with humor, games, stories, ads, backlinks, and click-bait. Sometimes I lose track of what I started to read because I literally fell into an Internet sinkhole and didn't even realize it. With high-speed Internet and the smartphone highway, there is no off-ramp.

REMEMBER

Saying no to something that everyone *does* and everyone *has* is hard. There is a feeling that you will somehow be left behind. Terms like FOMO (fear of missing out) and NOMOPHOBIA (fear of being detached from your mobile phone) speak to the idea that somehow, if you are not readily available with access to your screens, you will miss out on something important. Ironically, you have your eyes on your screens so much that you, in fact, miss *many* things that are important.

Striving for lower-tech (not no-tech) living

People live at a fast pace. Although many people are still not connected via high-speed Internet, Wi-Fi, and smartphones, a large percentage of the United States is. There are estimates that 81 percent of Americans have smartphones, with 96 percent owning a cellphone, and the numbers have gone up over 250 percent in a few short years, according to the Pew Research Center. Let's face it: People are wired, or rather wireless. Many people use their smartphones as their primary Internet access connection and are rarely without it, so essentially, they are never without the Internet.

The net effect of this (forgive the pun) is that you are always updating and receiving streamed input; in short, you have no downtime — ever. Although people are often online in some way, that doesn't mean they always have to be. The choice is now *internal* as opposed to *external,* not unlike food or nutrition, where you have a plethora of opportunities in terms of choosing your calories. However, people often choose foods based not on nutritional factors, but rather on taste, craving, emotions, and other reasons. At times, you may feel helpless in terms of your control over food choices, and the same may be true of your choices around Internet and screen access.

REMEMBER

Having the world at your immediate disposal and fingertips is powerful. It is also addictive and involves primitive reward circuits in the brain (see Chapter 2). This power needs to be respected for its potential impact on your life. Whether it is the desire and ability to video game, use social media, or just scroll around on your smartphone, all of it consumes your time and attention, and it can all lead to the stress of being *ever-on* — without periods of lower stimulation.

We have now become a *boredom-intolerant* culture. No one can be bored for even a moment. As an experiment, the next time you brew a pot of coffee or go into a waiting room, don't pull out your phone automatically. Just sit for a few minutes and do nothing. This almost sounds blasphemous. Do nothing? That means not be entertained by social media, TikTok, email, text, sports scores, banking, or stock trading, and not order those coffee filters that you forgot yesterday. Nothing. You may believe that this is a waste of time, but I contend that it is not. It is relearning the *value* of time. It is finding the nutritional value of *your* precious time that you've given to the mindless autopilot of your tech use.

REMEMBER

Life is lived in the gaps between tasks, not in the tasks themselves. Screens keep you busy, but not necessarily happy. The data regarding people who take a temporary or partial respite from screens unequivocally shows increased happiness scores and lower incidence of mental health issues.

Lower-tech living is exactly that: It is using technology and screens for what you need to, but cutting out the extra use that is harmful. The number that seems to work is about two hours a day, give or take. Numerous studies looked at this number, and it seems that less than that is too little and more is too much. Two hours is the *Goldilocks* number, although one study found one hour to be even better for children and teens. This is even more critical if you are under 25, when the frontal cortex of your brain is not fully developed, making you both more susceptible to addiction and having less executive judgment capacity.

Decreasing your stress with less tech

Whether you are an Internet addict or an over-user of your screen tech, the issue ultimately leads to the same result: diminished time and life imbalance. More subtle effects may also be associated with being ever-connected to your screens. Besides all the physical and psychological issues discussed in Chapter 10, people are unable to live fully in the present moment without distraction. In a sense, not only have they become *boredom intolerant,* but they have also become *distraction dependent.*

It might seem that I am overstating the obvious here, but using less tech means that you can spend that precious time doing the things you want and need to do. You may not realize what you have lost while you blindly donated your waking hours to your smartphone, laptop, and tablet, which all clamor for your attention. Your life erodes slowly, below your conscious awareness; then you find that you spent hours a day on your smartphone and all your other addictive screen activities.

The fact is that excess use of screen-based technologies increases your overall stress level. Some of this is due to increases in cortisol (the stress hormone), and some of it is simply the result of the amount of time that is spent. This leaves less time for everything else you must do, and that remaining time becomes compressed. When you compress life, you experience this as pressure and stress.

When they look at their screen use levels, most people are shocked to find how much time they are spending online. Even without an obvious addiction to video games, online pornography, social media, or other online activities, average users in the United States come in at 4 to 9 hours per day. Age seems to make a difference, with teens being in the highest use group. Think about this. Of the 16 waking hours we have each day, on average, approximately 25 to 50 percent of it is spent on nonproductive screen use. That includes TV, smartphones, tablets, computers, and so on. These numbers are staggering, but the data doesn't lie. Screen use is perhaps one of the few areas of your life that can be tracked, and it is one of the things you do that absorbs huge amounts of time, creating stress in its wake.

All the tech you consume is a constant call to action to your nervous system; this consumption is potentially limitless, but your capacity to manage the input and handle the psychological effects of excess screen use is limited.

Disrupting Your Tech Habits
with a Digital Detox

Sometimes it's necessary to limit access to an addictive behavior. Unlike most addictive substances and behaviors, screens are now so easily accessible that you may need to limit certain content and devices. Our clinic was one of the first outpatient centers in the country to utilize technology to help manage technology, and it allows us to help people in a way that they might not otherwise effectively receive.

To treat an addiction, patients need to have some period of at least partial detox, where they have an opportunity to reduce some of the established reward connections in the brain, and to help re-regulate their dopamine levels. Even for someone who overuses technology, placing limits on the total time or on problem websites or apps can help provide a safety net to limit their use.

REMEMBER

The Internet and screens aren't dangerous in and of themselves; rather, they are powerful and therefore can impact your brain, your behavior, and your life.

Defining a "digital detox"

REMEMBER

A *digital detox* is exactly what it sounds like: It is the absence, or in some cases reduction, of a substance or behavior. It results in a detoxification from the overuse of that substance or behavior. In this case I am talking about screens. A detox from the Internet and screens is different from a detox from drugs or alcohol in that users experience little physiological dependence and withdrawal. They experience some withdrawal symptoms, but these are typically seen in the form of psychological adjustment to reduced dopamine, and to the re-regulation of the brain's reward system. This period can be marked with anger, frustration, irritability, anxiety, depression, a feeling of loss, and overall discomfort.

Detox allows some of the well-worn neural pathways to degrade, thus allowing new behavioral patterns to be established; it also allows for the up-regulation of down-regulated dopamine receptors in the reward centers of the brain (see Chapter 2). The goal here is to break some of the biological and psychological connections with screen use.

Whether you're an addict yourself, or you have a loved one who is, or you simply want to be less connected to your screens, you need to change patterns of behavior. The behavior change sets up your nervous system to be receptive to new, rewarding behaviors and to allow you to obtain dopamine elevations from more constructive sources.

The use of *willpower* to manage your tech use is a bad bet. Everyone has good intentions to stop or limit behaviors, and Internet addicts and over-users will gladly promise themselves, and others, that they will change their behavior, but this change is generally short-lived. Not because they have no intention to do so, but rather because under a specific set of circumstances, willpower is not enough to prevent a habitual pattern from reoccurring.

Getting set for success

Many of the patients I see for consultation, parent coaching, or treatment at our clinic are under 30 years old. Most of the initial requests for help come from their parents or other loved ones. This presents some unique challenges in terms of supporting (and at times insisting on) changes to the screen-use habits of the child, teen, or young adult.

You cannot force anyone to do anything. Ultimately, you must clearly communicate your intentions and the appropriate consequences if they are not agreed on. Consistency, follow-through, and clear communication are necessary. It may also take good apps or software that can provide adequate blocks or limits and an IT consultation to help you structure any kind of limits at home. Most Internet addicts are incredibly resourceful when it comes to hacking through blocks, filters, or monitors placed on a device. I've seen Internet addicts find old computers and old smartphones that barely work just to get some access after parents removed their main screens. It's also my recommendation that none of this be attempted without an experienced mental health or addictions professional guiding you through this process. It is my experience that detox, absent of a comprehensive treatment plan, typically fails; the converse is also true, that treatment, without some limits and detox, also fails.

The Internet, smartphone, or video game addict will promise the moon when you threaten a complete or partial detox, but take care not to trust an addicted brain's promises. Note that the part of the brain that is involved with the addiction is not logical, and it only wants what it wants: to feel good. Your loved one will likely promise anything to get their screen back, but once they are reunited with their screen, the well-established pattern of addiction will quickly return.

Monitoring and Limiting Your Time and Content on Screens

Sometimes people decide to create a *dosing schedule* where they allow a more moderate amount of time online and on their devices. If there is an endless stream of potential access, the brain will want to use that time. When the resources become more limited, there is a natural adjustment to make do with what time is available. The following sections go over the basics of changing your time and content use online.

Turning off Internet access at a specified time

Time is everything when it comes to screen use, overuse, and addiction. It is the *fuel* that is used by the addiction. You cannot be a screen addict without spending a lot of time online and on your devices.

Time is the human capital that is usurped by screen technology, and it so happens that the Internet is very good at distorting your perception of time. So, what this means is that you must limit your own use of time (which is not always easy to do); if you're dealing with a loved one who is overusing or is addicted to their tech, then providing automated software or apps that limit total time consumption can be a very helpful approach.

TIP

In my experience, having these limits automated is key (ironic that I am advocating technology here) because it takes you, the parent or family member, out of the loop. In other words, the device is set to allow access only for a certain amount of time online, or it can be set to allow a certain amount of time on specific content areas such as video games, social media, or even adult sexual content. It is important to use reliable and easy-to-use software and applications; we at the clinic use Circle (https://meetcircle.com/) and Qustodio (www.qustodio.com/en/), but there are many other options as well.

REMEMBER

It's also important from our experience to have an IT person help set this all up — or at least guide you through the process. Many parents have some IT experience and want to do it all themselves, but we find this often does not work out well for a variety of reasons. Many complex emotions and family dynamics are involved, and if you are doing this for a child or teen, that can complicate the process of doing the technical setup. Having a parent do the setup also makes it that much easier for the addict to try to sidetrack the whole process and possibly try to manipulate changes in the treatment plan. Some parents handle it just fine, but many find the process difficult.

I have a tech (IT) person who works with our clinic and has had years of experience in setting up these systems. There is a lot of variation to each situation, so it's important to have someone who can be available for changes and adjustments, which are often necessary. If you want to find an IT person and the doctor or therapist you're working with does not have one, make sure you find one who is familiar with monitoring, blocking, and filtering programs like the two I mentioned earlier; make sure the IT person also interacts with the doctor or therapist who is coordinating the treatment plan. I want to also clarify that it is *not* my recommendation to only use software or applications to solve an addiction problem, but rather it is part of a comprehensive treatment plan to help the addict.

TIP

You can set a specific dose of Internet daily screen time, in addition to turning off all access to the Internet or specific content areas or apps. At the very least, I strongly suggest that devices, and perhaps the router itself, be set to go off at a reasonable time and/or after a specific dose of time for the addict. Left to their own devices (another pun), all over-users, and especially addicts, can easily be lulled into a *cybercoma* and stay on way longer, and much later, than they intended.

Limiting specific content

The idea of limiting specific content becomes even more critical in cases where there is an addiction to potent content. Again, the *synergistic amplifying* effect of the Internet modality, along with addictive and stimulating content, is a perfect combination for endless use. Blocking or limiting specific content can be invaluable in helping to treat a problem, while also preserving access to needed aspects of the Internet and screen use.

People must use the Internet. Too much of their lives are integrated with their smartphone and other screens to simply avoid them altogether. There are ways to block or limit complete content categories such as video gaming, gambling, pornography, social media, YouTube, and so on. Some of these content areas can largely be eliminated, while still preserving access to the necessary aspects of Internet and smartphone use; however, this can be tricky technically, so again, my advice is to get someone to help with the IT process.

One way you can limit content areas (websites or apps) is by creating an allowed list, called a *whitelist,* or a disallowed list, or *blacklist.* Sometimes you can combine both, depending on what you're trying to accomplish.

REMEMBER

Make sure that whatever approach you take with blocking, monitoring, or filtering, you do *not* give the addicted child *administrative privileges* for their devices; otherwise, they will be able to delete whatever software or app you install.

These issues can also become more complicated if your child is going to a boarding school or college, as you will have potentially fewer options; sometimes you may need to connect your IT person with the IT department at the school to coordinate planning.

REMEMBER

You'll likely have to tweak and adjust all these Internet blocks, limits, and filters several times. Sometimes they glitch, and often you'll need to adjust settings based on changing needs or other issues, such as equipment changes.

At times you may need more data to understand what is going on with your screen use or that of your child or loved one. In such cases, you may simply elect to monitor the time and content to ascertain where the problems are. Sometimes at our clinic we will use this step to collect information to help structure blocking and filtering, and sometimes it can be used to help educate the user or addict on how much time they are in fact spending and how they are using that time.

Establishing Values-Based Tech Use

Of all the suggestions I provide in this book, this one may ultimately prove the most useful. It may be harder to implement for a loved one, but in the bigger picture, this is where you must go to survive ever-encroaching screen tech in your life.

Values-based use is simply that. As you find out in this section, it is deciding how and when to use your screens (especially Internet-enabled devices) based on what your life goals and values are. The goal is to fit the screen technology around what you hold to be most important, as opposed to fitting in what is most important as an afterthought to your technology use. The reason is simple: If you use the latter approach, you'll never get to the most valued and important parts of your life. Internet technologies simply eat way too much time, and you often unconsciously allow such technology to dictate how you spend your time.

Introducing a values map

A *values map* is two things: a diagram that depicts how and where you spend your time, and a map that marks out your values. Unfortunately, they are often not aligned. The idea is to create a map that represents how you *currently* spend your time, and how you would *ideally* like to spend your time.

You need to see your technology from the perspective of how it can enhance your life, and not just because it is another new cool app, program, or website. I have many apps that I never use, and tons of junk email that I do not have time to get rid of. All of this consumes mental and physical energy, and even if a new app does something great, you need to ask whether it is good enough to justify the time it uses (including installing, maintaining, and learning it) and whether it takes away from other parts of balanced living.

A values map is based on the idea that you have limited time and limited attentional energy. You simply cannot manage numerous streams of information and content, so you must set *attentional boundaries.* That is a mouthful, but what you need to do with your screen time is what you do when you go through your junk mail: *get rid of what you do not want or really need.* By *really need,* I mean what will enhance your life, not simply be sort of fun or interesting. Be picky and do not let your time and attention be wasted without some scrutiny.

REMEMBER

Do not lie to yourself. And if it is a child or loved one who is overusing or is an addict, then try to help them to be honest with themselves. Mind you, you cannot make anyone see or do anything that they are not ready or willing to see or do. But that does not mean you cannot help structure life at home in a way that helps support a healthier and sustainable relationship with screens and the Internet.

Deciding where your eyes go

REMEMBER

By thinking about what is truly important to you — health, exercise, friendships, marriage or primary relationship, work, religious or spiritual activities, hobbies, travel, and so on — you can see what is needed to get these things done. The idea is not to use leftover time to live your life, but rather to use leftover time for entertainment, boredom management, avoidance of tasks, a dopamine hit, or just fun. Many things are fun, but they are not always as time-distorting as screens. You must decide where your eyes go. Don't let your eyes decide where you go.

A values map isn't actually a map; rather, it's a piece of paper where you track your thoughts and considerations (along with a conclusion and plan). I call it a map because it can help guide you through the process of making important choices. For example, if your main value is to stay connected to family and friends, do you keep Facebook and delete Twitter and other social media apps to focus on just one? Or do you delete all social media and focus on using FaceTime, Zoom, Skype, and so on to make direct calls and chats? The idea is to preferably use the most direct and efficient means to accomplish your goal to be connected to you friends and family.

The problem with social media is that it is a circuitous route to get to the desired goal of connection — and it is filled with ads, detours, distractions, and time-wasting entertainment. You end up spending your precious time watching cat videos on Facebook or Instagram. I know that if you're under 30, phone calling is less of a thing, but there really is no better way to connect (other than in person) with perhaps the exception of FaceTime, Zoom, or another video platform. Text your friend or loved one and set up a time to really connect, and save the social media for some limited entertainment if you must.

It is dangerous when it is so easy to waste several hours a day on your smartphone or on a video game. The problem with screens is that often you cannot easily manage time while you're on them, and it is this psychoactive and time-distorting power of screens that you need to address — *otherwise, a pastime will become all your time.* Values-based tech use acknowledges your limited time and energy; it also acknowledges the addictive allure of the Internet and your screens, and it assumes that life is what happens (at least much of the time) when you are *not* on your screen.

REMEMBER

Here's a phrase to consider: *Today is your future's past.* Sounds morbid, but it can be instructive. I often tell my patients to imagine how they would feel if they had very limited time left. What would they miss? What would they wish they had done more of? What would they want to say to certain people, and what websites, apps, games, or social media accounts would they want to log onto? Would you want to be posting on Facebook with limited time here on earth? Again, not to be morose, but *you do have limited time left.* We all do. Don't act like you have unlimited time such that you waste huge amounts of it online — yes, I said waste. To paraphrase Andy Dufresne in *The Shawshank Redemption,* you either "get busy living, or get busy dying" — there is nothing in between. Screens do not care if you are on them or not, so you must take charge of them — otherwise, they will take charge of you.

Removing Notifications and Addictive Apps

Sometimes it is necessary to remove an app, program, website, or notification. It is far easier to avoid overuse or addictive behavior if you do not have the most tempting lures in front of your face. If you are an alcoholic or watching what you eat, you probably do not want beer in the fridge or a box of chocolates in the cupboard. The idea is to remove the most tempting, problematic, and triggering components from your smartphone, computer, tablet, or another screen device. This is especially true of video games, social media sites, and even news sites that are purposely designed and gamified to keep you onscreen. This also goes for notifications, which are repeated invitations to engage in online behavior.

Knowing that notifications invite you to waste time

Notifications are a touchy subject. They can be a real problem on your smartphone because they act like the crack cocaine of smartphone apps; app developers and wireless providers are always asking you about notifications because they know this is a gateway to getting you onscreen. They know that if you receive a notification, you are more likely to use the app, thus keeping your eyes onscreen and perpetuating the addictive cycle and making money off you. Notifications trigger your brain's reward center and thus kindle the dopamine innervation cycle.

REMEMBER

The problem is that everything you have and do on your phone has a notification potential unless you turn them off or remove the app all together.

Getting rid of apps, websites, and software

Uninstalling apps can be complicated in that not only do you have to delete the app or program (or block a website), but you also must block the ability to download and reinstall it. It does you (or your loved one) no good to get rid of something on the smartphone or another device if you can just have it reappear from the cloud. It must be deleted completely, and ideally a block should be set up that will prevent the app or program from being reinstalled. This is more easily done on a smartphone or tablet, where you can easily set up a block to prevent the downloading of new apps; this can also be done on a computer, but it requires some external software or setting to accomplish this. An IT professional can help you with this.

WARNING

Do not assume that if you delete something addictive, either you or your loved one will not reinstall it in short order. I would say that unless you block the reinstall feature, it's highly likely that your willpower (or your loved one's) will quickly crumble, and back it will come to your smartphone or other screen devices.

Some blocking or monitoring software can notify whoever is keeping tabs on the software (the IT person, a parent, or the doctor or therapist who is directing the treatment) that there was an attempt to turn off the blocks or filters, or an attempt to reinstall something that was removed or turned off. This keeps everyone honest, and it is human nature (especially if a person is addicted) to push limits and to relapse. I mention in Chapters 1 and 2 that relapse is a very normal part of addiction treatment, and it's best to assume that this will occur and to take appropriate steps.

TIP

If you or your loved one are addicted, then it may be necessary to block the ability to download new or previous apps. Sometimes you can give the smartphone password to a friend or family member who can manage the blocks that are set up on the phone's operating system. Sometimes you may need other external software that I mention earlier, in which case an IT person would be helpful.

Filling Your Life with Real-Time Activities

Nature abhors a vacuum, and the same may be said for your screen use. If you stop addicted or heavy screen use, you need things to replace the dopamine hit you used to get and to fill the physical time gaps that are left open. A relapse to previous levels and pattern use is certain to occur if you do not help yourself or the addicted person to create a *real-time living plan.* The following sections can help.

The Real-Time 100 Living Plan

About ten years ago, I began to refine my theories and treatment techniques for Internet and technology addiction; I found that even if we could break the addictive cycle, there was still a need to quickly fill time with activities and behaviors that could help provide some of the dopaminergic innervation previously experienced from screen use. A window exists where the brain is looking for something that fits in that space. Your brain loves to organize, categorize, and make sense of your world — the idea is to give it something to fit into that empty hole.

REMEMBER

Left to its own devices (there is that pun again), the brain will look either to relapse or to find substitute stimulation to replace what was lost. The idea is to provide a reasonably benign substitute.

Watching hours of TV is a common activity that occurs when an addict stops excessive video gaming or other screen behaviors such as YouTube, social media, or even watching pornography. Switching from Internet behavior to watching hours of television is not what I would call a productive change, but sometimes we will allow this on a temporary basis as an interim step. However, I will in short order generally ask for a change to be made to decrease TV use, and that the patient (with our help) develop a list of 100 things that can be fun (and provide some stimulation), but that do not involve a screen of any type.

If you ask someone to write a list of 100 things that could be fun or interesting, they may initially say that there aren't 100 things on the planet that you can do without a screen. Take comfort that I have never had a patient who was ultimately unable to do this, although at times it took a few days of thought. It is important to note that you must do the *Real-Time 100* after you have had some detox (introduced earlier in this chapter); otherwise, there will be little to no motivation (remember, reward deficiency syndrome) to engage in developing the list, let alone doing any of the tasks.

REMEMBER

The idea is to create a list of 100 activities, behaviors, or tasks that you can easily do without involving a screen. (Figure 12-1 gets you started on your list; feel free to make copies of this figure and renumber the lines as you reach 100. Or, you can get out a pen and paper and make your list the good old-fashioned way.) Initially, you may feel there are not 100 things you can possibly do without a screen, but we have found that it just takes a little time to figure it out. The items should be relatively simple and easily doable. In other words, they should be things you can realistically do now, not in the future. You should also be able to do them with minimal equipment or supplies, although some basic materials are okay.

Real-Time 100 Living Plan

Instructions: The idea is to create a list of 100 activities, behaviors, or tasks you can easily do without involving a screen. Initially, you may feel there are not 100 things you can possibly do without a screen, but I have found that it just takes a little time to come up with them. The items should be relatively simple and easily doable. In other words, they should be things you can do now realistically, not in the future. You should also be able to do them with minimal equipment or supplies, although some basic materials are okay.

Name: _____ Date: _____

DOB: _____

1. _____
2. _____
3. _____
4. _____
5. _____
6. _____
7. _____
8. _____
9. _____
10. _____
11. _____
12. _____
13. _____
14. _____
15. _____
16. _____
17. _____
18. _____
19. _____
20. _____

FIGURE 12-1: With the Real-Time 100, you make a list of 100 activities that don't involve a screen. This sheet can help you get started.

TIP

Some tasks that our patients have used include hiking, listening to music, calling a friend, drawing, playing a sport of any kind, building something, taking music lessons or playing an instrument that you may already know, being outside in some way, flying a drone or model rocket, reading a book or magazine (preferably the paper kind), doing a puzzle, making a model of some kind, exercising, cooking a meal, teaching something you know to someone, going out for coffee or a snack, watching a movie with a friend or family member (yes, it's okay if you're doing it as part of a social experience), helping out a friend, and growing something.

Other self-help forms

The lists in this section are designed to help you better appreciate your Internet, video game, and screen use. We use these forms in our consultations, outpatient treatment, and residential treatment, and we find that they can help you or your loved one begin to look more closely at your (or their) relationship to screens. Many times, we do things unconsciously without thinking, and this automatic reaction keeps us stuck in habitual and addictive patterns. I suggest using as many of the lists as you can; you can use them to help you begin to develop awareness and to change your behavior patterns. These lists can also be used with the doctor or therapist you're working with.

What I like about the Internet, video gaming, or using screens:

1. _____
2. _____
3. _____
4. _____
5. _____
6. _____
7. _____
8. _____

What I feel in my body when I use screens, the Internet, or video games:

1. _____
2. _____
3. _____
4. _____

5. _____
6. _____
7. _____
8. _____

Emotion and feelings that increase my desire to use the Internet, screens, or video games:

1. _____
2. _____
3. _____
4. _____
5. _____
6. _____
7. _____
8. _____

What my Internet, video games, or screen use allows me to do better:

1. _____
2. _____
3. _____
4. _____
5. _____
6. _____
7. _____
8. _____

What my Internet, video game, or screen use stops me from doing or accomplishing:

1. _____
2. _____
3. _____

4. _____

5. _____

6. _____

7. _____

8. _____

What positives arise from using the Internet, screens, and video games:

1. _____

2. _____

3. _____

4. _____

5. _____

6. _____

7. _____

8. _____

What negatives arise from using the Internet, screens, and video games:

1. _____

2. _____

3. _____

4. _____

5. _____

6. _____

7. _____

8. _____

Personal life Goals — where I want to be in my life in five years:

1. _____

2. _____

3. _____

4. _____

5. _____
6. _____
7. _____
8. _____

Trigger list for Internet, video game, or screen use — a *trigger* **is something that makes you use (or want to use) your screen, Internet, or video game:**

1. _____
2. _____
3. _____
4. _____
5. _____
6. _____
7. _____
8. _____

Internet, video game, and screen use relapse prevention plan — a *relapse prevention plan* **is a written plan that outlines the things you can and will do to avoid relapse:**

1. _____
2. _____
3. _____
4. _____
5. _____
6. _____
7. _____
8. _____

Support groups, treatment resources, and organizations

There are numerous resources for dealing with Internet, screen, and video game addiction. The following lists reflect only a *partial sample* of some of the resources — including doctors, therapists, treatment programs, support groups, screen management apps and software, organizations, and information sources — that I have personally dealt with in my work. I am comfortable that the individuals and organizations in this section are offering quality content, information, and/or treatment services.

These lists are in no way meant to be comprehensive or exclusive, as there are numerous other options that are not included. My goal is to offer information and treatment resources that I am familiar with. This is not an endorsement, nor can I make a guarantee regarding their services. I received no compensation for the following listings. Information or healthcare services provided by those listed are the responsibility of the clinicians and organizations.

The treatment organizations that I am personally affiliated with are listed here:

>> **The Center for Internet and Technology Addiction** (www.virtual-addiction.com/) was initially founded in 1996 as the Center for Internet Studies, followed as the Center for Internet Behavior, by Dr. David Greenfield (yes, that's me). The center provides consultation, research, training, parent coaching and support, and treatment for Internet, video game, and technology addiction. We offer outpatient and intensive outpatient treatment (two- to four-week programs) and telehealth video services. Training, supervision, and workshops are available to healthcare professionals.

>> **The Greenfield Pathway for Video Game and Technology Addiction at Lifeskills South Florida** (www.greenfieldpathway.com) is a residential treatment program for Internet, video game, and technology addiction. It's designed and implemented by Dr. David Greenfield, who serves as consulting medical director. The program offers a four-phase residential treatment program for Internet, video game, social media, smartphone, pornography, and screen addictions. Treatment includes individual and group psychotherapy, medical and psychiatric consultation, family therapy, meditation and mindfulness, recreational therapy, exercise, eye movement desensitization and reprocessing (EMDR), and a variety of group and individual activities designed to help the addict return to balanced and sustainable use of technology, and to reclaim real-time living. Lifeskills South Florida also offers comprehensive psychiatric residential services for the treatment behavioral health problems and addiction disorders.

Resources and organizations

Resources, information, and organizations dealing with Internet, video game, and screen use, abuse, and treatment include the following:

» Al-Anon (https://al-anon.org/): Here you find 12-step support groups for friends and family of addicts.

» Center for Humane Technology (CHT; www.humanetech.com/): Co-founded by Tristan Harris, an ex-Google design ethicist, CHT is dedicated to radically reimagining digital infrastructure. CHT's mission is to drive a comprehensive shift toward humane technology that supports well-being, democracy, and shared information environment. Tristan was featured in the Netflix documentary *The Social Dilemma.*

» The Center for Internet Addiction (http://netaddiction.com/): Founded in 1995 as one of the first U.S. programs and websites for Internet and video game addiction by the late Dr. Kimberly Young. Dr. Young was an early pioneer in the research and study of Internet and technology addiction.

» Children and Screens: Institute of Digital Media and Child Development (www.childrenandscreens.com/): This group supports scientific research, public information, and parent guidelines on screen use and children.

» Circle (https://meetcircle.com/): Circle's parental controls let you manage screen time and monitor all websites and apps. It offers a complete in-home and on-the-go solution; you set the rules for sites like YouTube, TikTok, and other social media applications across all your family's connected devices. We use this program at our clinic and have found it quite effective and user-friendly.

» Common Sense Media (www.commonsensemedia.org/homepage): Find comprehensive information about screen use and media reviews for children and families.

» Digital Wellness Lab, Division of Adolescent and Young Adult Medicine, Boston Children's Hospital (https://digitalwellnesslab.org/): Dr. Michael Tappis and other clinicians offer services for treating Problematic Interactive Media Use (PIMU).

» Game Quitters (https://gamequitters.com/): Here you find information, resources, and support for gamers and their families.

» Sheila Hageman, MFA, MS (in progress) (https://everydaycreate.com/): Sheila offers coaching to parents around children's screen use.

» Healthy Gamer (www.healthygamer.gg/): The mental health platform created by Dr. Alok Kanoji is described as "able to help the Internet generation succeed." It helps offer accessible, inclusive, and affordable mental health resources that empower the Internet generation to find peace and purpose through content, community, and coaching.

- » Internet and Technology Addicts Anonymous (ITAA; https://internet addictsanonymous.org/): Here you find 12-step support groups for Internet and technology addiction.

- » Lindner Center of HOPE (https://lindnercenterofhope.org/what-we-treat/addictive-disorders/): Lindner Center of HOPE offers screening and treatment options for Internet and video game addiction, as well as other behavioral and substance-based addictions.

- » On-line Gamers Anonymous (www.olganon.org/home): This is a 12-step support group for video game addicts.

- » Qustodio (www.qustodio.com/en/): Qustodio offers parents and loved ones tools to help monitor, supervise, and control access to the Internet and screen devices. We use this program and have found it quite effective and very robust in what it can do. I have found that using an IT person with this program can be helpful, although it can be done by anyone with basic computer knowledge.

- » Restart (https://restartrecovery.org/): Restart Recovery is the first residential treatment program in the United States to offer treatment for Internet addiction and related disorders. They offer comprehensive behavioral addiction services including Internet, video game, and social media addictions, as well as other behavioral addiction treatments. They serve children and adolescents as well young adults.

Doctors and therapists

The following doctors and therapists offer a variety of outpatient treatment services for Internet, video game, and technology addiction:

- » Dr. Mike Brooks (www.drmikebrooks.com/)

- » Dr. Don Grant (doctordeeg@gmail.com)

- » Dr. Christine Miller (www.indianahealthgroup.com/provider/christine-miller/)

- » Melissa Rahming, LDAC (melissa.rahming127@gmail.com)

- » Michael Shelby, MA, LPC (https://technologyaddictioncenter.com/)

- » Dr. Ed Spector (www.spectortherapy.com/wp/)

- » Dr. Cliff Sussman (https://cliffordsussmanmd.com/)

- » Traverse Counseling and Consultation (https://traversecc.org/)

Additional links

Additional resources and links include the following:

- A video on what causes addiction: (www.youtube.com/watch?v=C8AHODc6phg&t=72s)

- American Academy of Child and Adolescent Psychiatry (www.aacap.org/AACAP/Families_and_Youth/Facts_for_Families/FFF-Guide/Children-And-Watching-TV-054.aspx)

- American Academy of Pediatrics

  - https://services.aap.org/en/search/?context=all&k=screen%20time

  - www.healthychildren.org/English/family-life/Media/Pages/Where-We-Stand-TV-Viewing-Time.aspx

- American Psychiatric Association (www.psychiatry.org)

- American Psychological Association

  - www.apa.org/topics/social-media-internet/technology-use-children

  - www.apa.org/pi/families/resources/newsletter/2019/05/media-use-childhood

- American Society of Addiction Medicine (www.asam.org/quality-practice/definition-of-addiction)

- The Center for Parenting Education (https://centerforparentingeducation.org/library-of-articles/kids-and-technology/how-much-time-internet-kids/)

- Children's Screen Time Action Network (https://screentimenetwork.org/about/leadership-team)

- KidsHealth (https://kidshealth.org/en/parents/net-safety.html)

- Massachusetts General Hospital (www.massgeneral.org/children/internet-addiction)

- Mindful Schools (www.mindfulschools.org/inspiration/addicted-to-technology-how-to-stop-panicking-and-create-healthy-solutions)

- Norton Family (https://us.norton.com/norton-family)

- Pew Research Center (www.pewresearch.org/internet/2020/07/28/parenting-children-in-the-age-of-screens/)

- » Planned Parenthood (www.plannedparenthood.org/planned-parenthood-northern-new-england/parents/internet-safety-tips-parents-teens?)

- » Psychology Today (www.psychologytoday.com/us/archive?search=internet+addiction&section=All)

- » SAMHSA National Helpline (www.samhsa.gov)

- » Science Direct (www.sciencedirect.com/science/article/abs/pii/S0360131511000273)

- » Screen Time (https://screentimelabs.com/)

- » *This Panda Is Dancing* (https://www.youtube.com/watch?v=tf9ZhU7zF8s)

- » WebMD (www.webmd.com/mental-health/addiction/video-game-addiction)

- » World Health Organization (WHO; www.who.int/news-room/q-a-detail/addictive-behaviours-gaming-disorder)

- » Zuckerberg Media (www.zuckerbergmedia.com)

IN THIS CHAPTER

» **Understanding addiction terminology and psychotherapies**

» **Deciding which type of treatment, if any, is best**

» **Determining motivation and readiness for treatment**

» **Comparing various treatment options**

» **Treating co-occurring psychiatric issues**

Chapter **13**

Exploring and Evaluating Treatment Options

Treatment for Internet and video game addiction is a relatively new and emerging area of addiction medicine. When I started treating this issue in the mid-1990s, there was essentially no research on evidence-based treatment options for Internet addiction. Today, hundreds of studies exist that have examined which options might be suitable for the treatment of Internet, video game, and screen-based addictive behaviors. In this chapter, I review some of the highlights of the treatments that I have found to be effective.

REMEMBER

Addictions are often initial (albeit counterproductive) attempts to solve a problem in a person's life. To that end, addictions often work, at least for a while, but then they often end up *becoming a problem unto themselves.*

TECHNICAL STUFF

A word about case-based, anecdotal data versus research-based data: Although much of this book is based on my 25 years of clinical work with Internet and technology addiction, some of the data this book is gleaned from various research studies I have utilized to inform our treatment process. When it comes to a patient walking into our offices, however, we must base what we do on what we believe will help; sometimes these treatments jibe with what the evidence-based research suggests, and sometimes they do not. There isn't always a congruence between sample-based research and the actual patient (or family) care.

First Things First: Talking about Terminology

There is a problem in clearly defining what we're treating. As I discuss in Chapter 10, numerous diagnostic labels are associated with Internet and screen behaviors — including Internet addiction, video game addiction, Internet gaming disorder, pathological Internet use, problem Internet use, problem Internet and media use, and Internet use disorder — and other screen-based addiction labels. If this sounds confusing, that is understandable.

REMEMBER

At this point, we're still refining the diagnostic definitions enough to be able to speak the same language when we're describing a complex pattern of behaviors such as Internet addiction. It's quite possible that in the future, we will further refine Internet and screen addiction into several diagnostic categories; for the purposes of this chapter, I'm describing a general issue involving Internet and technology addiction that will subsume video game addiction, social media addiction, and online pornography addiction, as well as general overuse of smartphones and other Internet-based screen technologies. All these examples share a common feature: excessive screen use to a point where there is notable impact on one's well-being functioning.

From the early days of addressing this issue, I've always taken a pragmatic approach to helping my patients. In addiction medicine, clinical psychology, and psychiatry, we are charged with helping the people who come to our offices, even if we do not always have well-defined definitions for the problems being treated, and this was initially the case with Internet addiction and screen-use problems. Beginning with video game addiction and expanding to problems with pornography, social media, smartphone use, and excessive Internet use, we began to see screen use as a growing clinical issue. In this chapter, I rely on my reviews of the medical literature, as well as my clinical experience over the last 25 years treating Internet addiction and related disorders.

Much of the information discussed in this chapter and throughout this book comes from my years of clinical practice in diagnosing and treating individuals with Internet and screen addictions. Although I'm familiar with the medical literature, research, and scientific evidence, some of the conclusions that I have made come from my day-to-day treatment of these patients. When I started treating people who were showing up at my door in the mid-1990s, we really didn't have a lot of research to go on. As a practicing addiction medicine psychologist, I have conducted research, written, and published in academic medical journals, and trained hundreds of doctors and therapists in treating this issue; a large part of my expertise comes from having spent thousands of hours with patients (and families) who have great difficulty in living life without their screens. When it comes to helping patients (and their loved ones), necessity often becomes the mother of invention.

Checking Out Different Psychotherapies

Addiction treatment is not a simple process of plugging in one specific treatment approach, as there are often numerous dimensions to treating and managing this complex problem; for instance, we see many life issues because of excessive screen time, including sleep problems, lack of physical exercise, weight gain, depression, anxiety, and lack of motivation for real-time living, to name but a few. Chapter 10 outlines these consequences in greater detail.

Psychotherapy (of various types) is one tool that is used in the treatment of Internet and screen disorders, and typically, many of our patients have received counseling or psychotherapy prior to receiving treatment at our clinic. Psychotherapy alone only treats *part* of the addiction problem, but Internet and video game addiction is a complex issue that often requires management on several fronts.

REMEMBER

What is psychotherapy? *Psychotherapy* is the scientific application of certain psychological techniques and principles to assist a patient in understanding, managing, and changing behavior. Good psychotherapy is not simply about insight or understanding, but must also include aspects of addressing behavior change. Decades of research show that psychotherapy is a highly effective treatment for a variety of psychiatric disorders and conditions, but when it comes to addiction, psychotherapy becomes only *one* tool in the overall treatment process along with other modalities. The following sections go over different types of psychotherapies and other therapeutic treatments used in treating Internet and technology addiction.

Using cognitive behavioral therapy

Over the last 40 years, cognitive behavioral therapy (CBT), along with its many offshoots, has become one of the most researched and widely used types of psychotherapy.

CBT examines patterns of thought, emotion, and behavior that contribute to, and maintain, a person's symptoms. This may sound simple, but people are often largely *unaware* of how their thoughts contribute to their behavior and how their perceptions contribute to cycles of problematic addiction. CBT looks at the interaction of the addict's thoughts and subsequent behavior and how this interaction impacts addictive behavior and other psychiatric issues.

REMEMBER

More evidence-based research likely exists on the use of CBT in Internet and technology addiction treatment than for any other subtype of psychotherapy. CBT helps examine irrational and destructive thoughts that contribute to excessive screen behavior. The idea is to deconstruct patterns of thought (along with emotional reactions) that connect to both underlying psychological pain and conflict, as well as resultant psychological symptoms created by an addictive behavior pattern.

Understanding medication-assisted therapy

Medications can be a useful adjunct in the treatment of Internet and technology addiction. Although medications can help in the management of addiction, they are often more useful in addressing some of the co-occurring or resultant conditions that go along with screen-based addictions. Common diagnoses include depression, social and generalized anxiety, obsessive-compulsive disorders, Autism Spectrum Disorder (ASD), and the most common diagnoses seen with Internet and video game addiction — ADHD.

It is important to be neither afraid of the use of medications (as they can be quite useful and effective), nor overly reliant on the expectation that they will make the ultimate difference. There is a growing acceptance and good scientific support for the use of medication in addiction treatment, and some studies speak to the specific use of medications with Internet and video game addiction.

REMEMBER

The term *medication-assisted* means just that. No pill of any kind can solve or cure the complex biobehavioral problem of Internet addiction. Medications are another tool in treating the totality of symptoms and are a part of a complete treatment plan. If you're using a separate clinician (physician, nurse practitioner, prescribing psychologist, or physician's assistant) to provide medications, make sure that all the providers involved in your care, or your loved one's care, are communicating with each other. This can be a challenge logistically, but overall, it will ensure a more positive outcome if all caregivers are communicating.

Medications that may be useful in treating Internet addiction

Numerous studies have looked at the effectiveness of medication in treating Internet and video game addictions. Rather than go through each study, I'll provide a basic synopsis of the highlights. Many of the studies used a small number of subjects, and some were *anecdotal* or *clinical* in that they were not blinded or experimentally controlled. Some of the terminology may be unfamiliar, but bear with me:

>> Alpha and beta-adrenergic blockers can be used to treat social anxiety and generalized anxiety, and some of the medications used include propranolol, clonidine, trazodone, and prazosin. At times, selective serotonin reuptake inhibitor (SSRI) antidepressants have been used for their anti-anxiety and anti-compulsive properties with some success, although my experience suggests a less-than-ideal response from SSRIs for Internet and screen-based addictions. In general, addiction and impulsive disorder treatment with SSRIs seems to be most effective when obsessive-compulsive features are present. Keep in mind that Internet addiction involves both *impulsive* and *compulsive* behaviors.

>> Some positive results have been seen at our clinic with the use of bupropion to reduce cravings for Internet video gaming, but only a few studies support this. Some neuroimaging studies show decreased activity in the prefrontal and parahippocampal parts of the brain, although correlations between functional imaging studies and subsequent clinical effectiveness are not always found.

WARNING

>> Mood stabilizers such as carbamazepine, gabapentin, and lamotrigine are not well studied, and we have not used them very often, unless there were other psychiatric reasons to do so. I would say overall that they are not specifically useful for managing Internet and technology addiction but can be useful if significant mood dysregulation is present. We *have* seen some effectiveness in using gabapentin for anxiety management.

WARNING

>> Opioid receptor antagonists (naltrexone) — borrowed from success with pathological gambling, alcohol abuse, and substance use disorders — have shown some success, but little specific research has been conducted with screen-based addictions. The logic for the use of these drugs makes sense in that they impact the dopaminergic pathways, inhibiting dopamine release in the nucleus accumbens, thus interfering with the neurobiological reward process. Some patients do not tolerate these drugs well, so if they are used, they should be started at a low dose and increased slowly.

>> Psychostimulants are very commonly used for concomitant ADHD, with Concerta, Focalin, Ritalin, Adderall, and others being very useful in managing typical ADHD symptoms. They do not, however, seem to directly impact the addictive process, although theoretically they should. They do manage some of the *preexisting* and possible *resulting* attentional issues that often co-occur with Internet screen disorders. There may be some indication that ADHD symptoms are increased from excessive screen use, and these drugs may

mitigate some of these effects. It has been my experience that many patients who have ADHD are using their screens to self-medicate their ADHD symptoms, but it is quite possible that they are making them worse by further desensitizing themselves to normal levels of stimuli.

>> The use of antipsychotics is less common, although some evidence suggests efficacy with the obsessive-compulsive disorder (OCD) and impulse control components of Internet gaming addiction. In our residential program, we have used these medications to help manage some aggressive agitation and mood dysregulation during the detox phases of treatment, but their use is often temporary, unless other psychiatric symptoms are present.

A word about treating other psychiatric conditions simultaneously

REMEMBER

Overall, medications are more typically useful when there are co-morbid or co-occurring disorders, but not as the primary therapy for Internet addiction and related disorders. Medication-assisted therapy (MAT) still has few evidence-based treatments for Internet addiction, but more research is on the way, and ultimately, these are clinical decisions made by those treating you or your loved one.

WARNING

It's important to emphasize that addictions often create secondary symptoms because of life imbalances from excessive screen time. I frequently see depression resulting from an addiction, but some of the depressive symptoms may be from reward deficiency syndrome and a reaction to functionally depleted levels of dopamine. Dopamine desensitization makes real-time living feel less rewarding, and this may be experienced as depressive symptoms (see Chapter 2). The interesting thing about this type of depression is that it seems to be limited to all things *except those online.* In other words, screens carry significant positive reward potential for dopamine innervation, but real life does not. In such cases, antidepressant therapy may do little to improve the net mood (no pun intended) of yourself or your loved one.

Trying family support groups and therapy

There is perhaps no greater example of the need for family therapy than in the case of an addiction — addictions of all types impact everyone in the home and family. The actions and behaviors of an Internet addict affect other family members because behavior changes, including increased isolation, mood dysregulation, and avoidance of real-time living. These changes can disrupt family functioning, and more importantly, they can create serious concern from parents and loved ones for the health and welfare of their family member.

Disruption of family routine can often occur, even if the addict doesn't live with the parents or family or have knowledge of the ongoing addictive behavior and its consequences. For example, reduced grades, missing work, social isolation, depression, poor hygiene and self-care, and weight gain (or loss) can result in many a night of lost sleep for a parent or loved one. Watching a loved one's life decline is painful and scary, and many parents I speak to feel helpless, hopeless, and scared (as well as frustrated).

The following sections cover the role of family and loved ones in Internet and video game addiction treatment.

A reversal of roles

Typically, parents know more than their children, but not when it comes to the Internet and screen technology. The last two-plus generations have been born and raised with the Internet and screens and, frankly, have an almost intuitive grasp of how computers, the Internet, and all things digital work; this poses a unique problem in trying to set limits in the home (which becomes even more complex if the child is away at school). I cannot count how many times I've been told by a parent that they set up blocks, limits, filters, or monitors and that they were effortlessly disabled by the addict.

REMEMBER

Unfortunately, this means that parents and loved ones must become much more knowledgeable, and if they cannot learn the technology, it is critical that they arrange for a competent IT professional to help. If you're working with a doctor or therapist who is an expert in Internet and technology addiction, they should be able to refer you to a competent IT professional with experience in this area. If you have trouble with this, you can always contact our clinic and we may be able to help as well. Obviously, ask for references, and most importantly, ask how many people they have helped with this problem. Even if a parent is computer savvy, they may not have the level of technical sophistication or emotional stamina needed to do it alone, as the emotions that are involved in dealing with a loved one make objectivity difficult.

The doctor or therapist you're working with can help you decide what types of therapy, treatments, and interventions are best — including medications, IT blocking and filtering, various psychotherapies, eye movement desensitization reprocessing (EMDR), and support groups — or if residential treatment is indicated. Some of these decisions are dynamic, meaning that they can change throughout the treatment process as more information becomes known. I would not recommend making all these decisions as a parent or loved one without some professional guidance and support.

Clear, strong boundaries

Learning to maintain healthy boundaries with an addict is critical. Addictions can impact everyone at home, and this can spill out to everyone in the family. Care should be taken to set appropriate limits and to be very clear what you, as parents or loved ones, can and will do, and what you will not do. Being indecisive helps no one. Boundaries are just that: They are limits and they need to be expressed in a clear and calm manner. Being clear with your boundaries does not mean you are short or angry — just clear. If you, as a family member, get angry or frustrated when dealing with an addict, then the addict knows they have *some* control the situation, and this easily allows the addict to dismiss emotions of family.

REMEMBER

An addict (especially if they are under 25) has much more time and energy than you as a parent or loved one. They know all your vulnerabilities and how to push your buttons (and in many cases, they installed those buttons). They have been watching you all their life. The less you react emotionally, the more power and objectivity you will have in helping your addicted loved one, but this is not easily done.

Family support groups

TIP

I mention support groups in Chapter 12; the one I like best is Al-Anon (https://al-anon.org/). Al-Anon is a support group for friends, family members, and loved ones of addicts — not specifically Internet and video game addiction, but for all addictions. Although the parents and family members who go to these groups (both in person and virtually) are dealing with all types of addictions in their loved ones, the bottom line is the same. The strength, hope, and encouragement that come from these meetings are invaluable. They can make an enormous difference in learning to set limits and boundaries, and in learning to separate when necessary from a loving and compassionate place. I cannot overemphasize how helpful these meetings can be for family and loved ones who are dealing with an addiction of any kind, although you may be surprised to find that a large part of the support you receive is around setting good boundaries and taking care of yourself, and not just the addict.

Family therapy

In my experience, utilizing family therapy as a part of a treatment plan is essential; this is especially true in residential treatment (covered later in this chapter). At our residential program at the Greenfield Pathway for Video Game and Technology Addiction at Lifeskills South Florida, family therapy is mandatory. Sometimes this therapy involves the parents alone, and at other times it involves the addict and family members.

The goal of family treatment is to help empower parents to set limits, rules, and consequences on a consistent basis, which frankly can be very difficult in these situations. Many times, parents have provided all the technology and high-speed Internet for their children. This is always done out of love, but ultimately it can be a contributing factor to the development and maintenance of an Internet and technology addiction — therefore, it becomes an expectation of the addict that they will continue to be provided with the equipment and services. Other times the family therapy is focused on other interrelationships and dynamics in the family that may be contributing to or maintaining the addiction.

REMEMBER

One final note about family therapy: Most addictions are family problems. They involve and affect the whole family, and often create complex dynamics and interactions between family members — as well as with most parents. Family therapy can be a critical ingredient in addressing disagreements between parents or other family members, and can be a significant part of the equation in helping the addict recover.

Focusing on group therapies

Typically, outpatient treatment consists of a doctor (psychologist or psychiatrist) or therapist working with the addict on a one-to-one basis, providing psycho-therapy, medication, and family therapy. Group therapy may or may not be utilized, although there are some exceptions if the doctor or therapist has enough patients dealing with similar issues to offer group therapy treatment or can refer to another clinician nearby who has a group running. However, if you're using an intensive outpatient program, partial hospital program, or residential treatment program, then group therapy will usually be a very large part of the treatment.

REMEMBER

Group therapy may be one of the most effective forms of treatment for addiction; although it's not always practical to do on an outpatient basis, it can be more eas-ily utilized in a residential setting. Groups use the power of social interaction, peer bonding, social comparison, and social skills facilitation to help create change. The group forms a bond and a set of norms that act as both a catalyst to change as well as a support. A good group can be confrontational as well as understanding, and therein lies its power.

The other factor that makes a group work so well is that typically all the members of the group have a similar issue. In this case I'm talking about Internet and screen addiction. If you have six young men in a group who all have a video game or Internet addiction, this similarity becomes a basis for connection, comparison, and understanding.

Some groups are focused on a topic or theme and are designed to address a spe-cific issue. Often residential programs have numerous focused topic groups, such as social skills, tech wellness, relapse prevention, assertiveness, meditation/mindfulness, and other topics. The idea is that the group provides a specific set of skills, knowledge, and practice to help the addict when they are in different situations.

Figuring Out Which Treatment Is Best

Deciding on a type of treatment is not an easy task. For the parent or loved one, how do they know which clinician is trained and experienced in Internet addiction treatment? How can they tell who is the best doctor or therapist to help their loved one? And how do they know what type of treatment is best? I do my best to answer these questions in this section.

First, and perhaps foremost, you need to assess the credentials and experience level of the psychologist, psychiatrist, or therapist you're dealing with. The behavioral health and addictions field is complicated, and numerous professions, degrees, certifications, and licenses often provide the same or similar services. Many types of providers can, and do, offer the same or similar clinical services; their training might be different, but their scope of practice according to their state licenses may have some overlap.

TIP

The best way to know is to ask a lot of questions; ask the provider about their training, experience, treatment philosophy, and practice scope. The provider's answers should help guide you. Also, ask around, talk to your primary care provider, search online (yes, that's a good use of the Internet), or ask other mental health or addiction clinicians who they might recommend for the treatment of Internet and technology addiction. In each community, usually a few names will emerge, or with telehealth and remote treatment options, you may be able to access providers outside your local area.

One note: You may also need to ask about costs and insurance participation — some doctors and therapists do not participate in insurance networks but may support out-of-network services.

REMEMBER

Perhaps it's less critical whether they have a PhD, PsyD, MD, DO, APRN, NP, or other clinical master's degree of some kind, as compared to the level of experience and expertise they have with regularly treating Internet and technology addiction. Many doctors and therapists have experience treating addiction, but few of them specifically treat Internet and technology addiction. *It is critical that you ask how many patients they have treated and how long they have been treating these problems.* Unfortunately, addiction medicine and behavioral health is like any other business, and sometimes doctors and therapists (and this goes for residential treatment programs as well) will say that they treat something, when in fact they really do not have a lot of experience with this population. Technically, this is not necessarily inaccurate, because once licensed to practice, they are theoretically allowed to treat many diagnoses, but ethically, they should only treat those problems that they have the experience and training to treat. Do not be afraid to ask questions, and if they seem evasive, defensive, or unwilling to answer, then consider going somewhere else.

Internet and technology addiction is fast developing as an emerging specialty in the addiction medicine sub-field of process and behavioral addiction. Although there are similarities to other forms of addiction (especially on a neurobiological level), there are some unique differences that do require specialized experience and skills. It's perfectly appropriate to ask about the doctor or therapist's training and experience with the type of screen issues you or your loved one is having.

REMEMBER

If you're comparing doctors, therapists, or residential treatment programs, then you really need to make sure you are accurately comparing things that are the same. Traditional outpatient treatment is not the same as an intensive outpatient program or, for that matter, a residential program. You need to compare similar programs and services to make an accurate comparison.

REMEMBER

No indicators are perfect in determining what level of care is right for you or your loved one. Generally, although not always, we start with the least restrictive (or simplest) level of care. Most of the time, this starts as outpatient treatment (either in person or via telehealth), and then, if the response is not as desired, we might move up to an intensive outpatient or partial hospital program; lastly, we might go with a residential program where you would live at the facility for a period of time (typically 30 to 90 days, although sometimes longer). I cover these programs later in this chapter.

Our program at the Greenfield Pathway for Video Game and Technology Addiction is designed to be a 90-day program, but sometimes people stay for shorter or longer periods depending on their specific needs and the other co-occurring psychiatric issues that need to be addressed. Some programs offer step-down options, where you or your loved one would live at a semi-independent home in the community while still receiving services at the residential treatment center as well; this approach supports the reintroduction to real-time living and dealing with the more challenging aspects found in day-to-day life (including technology). Sometimes a patient may start at one level of care and move up or down, depending on their response or their needs. It is not a one-size-fits-all approach, as individual differences often need to be considered.

When it comes to addiction, it's often necessary to have a period of detox where a patient has *limited* or *no access* to technology. The detox period is critical, as is the ability to have distraction-free treatment where you are not spending 12 hours a day on your device. The only definitive way to accomplish this is in a residential format. We do treat patients in our outpatient and intensive outpatient programs and can do a partial detox if we have a good set of monitoring, blocks, and controls on their devices and technology. This is not always easy, but it allows us to effectively treat and manage Internet addiction when they are living at home or at school.

REMEMBER

Sometimes it's necessary to use several doctors or therapists to treat yourself or a loved one. One doctor or therapist may conduct a part of the treatment, while another might focus on medication, and another might run family or group treatment. It is not unusual to use multiple clinicians to help make the treatment effective and practical. In a residential program, numerous clinicians will typically be working with the addict as well as the family.

Prepping for Treatment

Perhaps as important as what type of treatment is how to know when you or loved one is ready, willing, and able to participate in *any* kind of treatment. Readiness and motivation are crucial factors in determining if, and what type, of treatment you or loved one should or could attend. The idea is to access what might work best and when to start, and there are no perfect formulas for deciding. Sometimes it is a matter of logistics or geographic location; there are also financial and insurance considerations, and timing in terms of school, jobs, and personal commitments all must be weighted in the equation. Lastly, of course, is the addict's willingness to participate.

Surveying motivational enhancement and harm reduction strategies

Motivational enhancement is a technique often used in addiction treatment where the doctor or therapist uses *motivational interviewing* techniques designed to elicit information, increase the patient's openness, and slowly increase their motivation and willingness to participate in treatment. This is half art and half science, because most of the time, the patients we see who have an Internet or technology addiction do not necessarily think they have a problem. To some extent, they may be in denial and simply think their parents or loved ones are mistaken, or they do not understand the impact of their excessive screen time. Many Internet and technology addicts we treat believe they are being productive (although that is not actually true), even with their excessive time online.

Motivational interviewing is an effective way to counsel and help plant the seeds of change in someone who is unmotivated or ambivalent about changing their addictive pattern. While no one chooses to be an addict of any kind, once they are in the neurobiological and behavioral cycle of addiction, they are often unable to appreciate the negative consequences of an addiction that may be occurring. Motivational techniques meet the addict where they are, and help (in a subtle and gentle manner) increase their insight and potential motivation to change. This is not always successful, but in my 30-plus years of treating general addiction, I've found these techniques to be invaluable in helping addicts of all types, and specifically Internet and technology addiction patients.

Motivational interviewing generally involves open-ended statements to the addict about their life. Making empathetic comments to the patient, reflecting frustration with their situation (which may include how upsetting it must be to have their parents on their case), helps motivate the patient. The idea is to help the patient see that you're aware of their perspective, and that you are not just another person who is going to tell them what to do. Avoiding arguments and confrontation with the patient helps enlist them in the treatment process and minimizes their defensiveness.

Harm reduction is simply the idea that *some change is better than no change.* Addiction treatment is often progressive and cumulative. Sometimes the best we can do is to make small inroads of change that build over time. This is not necessarily a failure but is rather built on the idea that some change now leads to more change later. If you have an Internet addict who is going online to game, surf/scroll, TikTok, or YouTube 12 hours a day, and you can reduce that to 6 hours a day, that is a form of harm reduction. When it comes to screen addiction, some change is always better than no change, but abstinence from specific addictive content and overall reduction of screen time may still be the ideal goal.

REMEMBER

Treatment of addiction is often cumulative and may require several attempts. Relapse is not a failure, but often is part of the treatment process, and you need to look at treating an Internet and technology addiction as a *recovery process,* as opposed to a one-time event. Our greatest gift in medicine as clinicians is the *installation of hope.* Sometimes it's difficult for us to feel hopeful or to adequately convey it, especially when we see our patients not getting well or immediately relapsing. Medical compliance and treatment adherence, across all illnesses, is about 50 percent; yet in the treatment of addictions, we somehow expect better or higher compliance than in general medical care, and when we don't see it, we might interpret this as a lack of motivation or commitment.

Assessing readiness for change

Readiness for change is a system of categorization for a person's readiness to change some aspect of their life. It's often used in staging a patient's readiness for behavioral health or addiction treatment. I've found this system to be extraordinarily useful in assessing the readiness of patients to enter treatment and to gauge their overall level of desired change, motivation, and willingness to make life alterations.

In a journal article published in 1992, psychologists James O. Prochaska, Carlo C. DiClemente, and John C. Norcross detailed the stages of readiness for change and the matched interventions as follows:

>> **Precontemplation:** Not currently considering change — "Ignorance is bliss."

Validate the addict's lack of readiness. Clarify that the decision is theirs and encourage regular reevaluation of their current behavior; encourage exploration, not action. Explain the pros and cons of change, and personalize the risk to their specific situation and circumstance.

>> **Contemplation:** Ambivalent about change — "Sitting on the fence."

The addict is not considering change within the next month. Validate their lack of readiness. Again, clarify that the decision is theirs. Encourage evaluation of pros and cons of behavior change; identify and promote new, positive outcome expectations.

>> **Preparation:** Having some experience with change and trying to change some aspects — "Testing the waters."

The addict is planning to act within one month. Identify and assist in problem-solving regarding any potential obstacles. Help the patient identify social supports. Verify that the patient has underlying skills for behavior change; encourage small initial steps.

>> **Action:** Practicing new behavior for three to six months — "Early changes."

Focus on restructuring cues and social support. Bolster self-efficacy for dealing with obstacles; combat any feelings of loss, and reiterate long-term benefits. Be supportive of the change process.

>> **Maintenance:** Continued commitment to sustaining new behavior post-six months to five years — "Mature changes."

Plan for follow-up support. Reinforce internal rewards. Discuss coping with a relapse if it happens and reiterate that relapse is not failure.

>> **Relapse/Resumption:** Resuming old behaviors and patterns: "Fall from grace."

Evaluate triggers for relapse. Reassess motivation and barriers. Plan for stronger coping strategies. Use compassion for self and provide support. Be careful not to make judgments.

REMEMBER

Many changes that people make in their lives (not only medical ones) can be understood through a readiness-for-change analysis. In a sense, we must all move through various cognitive and emotional processes that allow us to engage in the difficult (but often necessary) process of change. It's also important to clarify that these are not necessarily linear and progressive stages; one can move back and forth through several stages and revisit earlier stages, especially when there is a setback or relapse. Sometimes, when it comes to treatment, you must dip your toe in the water first and then take some time before you do it again. The idea is to understand that change is a complex process that is progressive in nature.

Using Alternative Interventions and Other Treatment Approaches

There are many ways to help someone with an Internet or video game addiction. Sometimes there are differences in preferred treatment options because of the online drug of choice. For example, a video game addiction might be more complicated to treat or manage, in part because a video game can be accessed in many ways. You can choose from free-standing console game devices (such as PlayStation and Xbox), handheld game devices (like Switch), and still more options in terms of PC-based gaming that can be played on a fast laptop or desktop computer. And lastly, there is the growing use of smartphone gaming, which has almost become the crack cocaine of gaming for many people who become hopelessly addicted to their smartphone games. The amazingly easy access of a smartphone game allows people to play anytime and anywhere, and this *lowered threshold to act* is a significant risk factor in all addictive behaviors.

As I explain earlier in this chapter, the research literature on the diagnosis and treatment of Internet addiction and screen use disorders is new. We have been actively studying these issues only for the last 25 years, and there is a lack of evidence-based data about which treatments might be best for different types of Internet addiction disorders. We do know a fair amount about other aspects of addiction medicine, and we have developed some overall treatment strategies in terms of the types of treatment that are helpful in screen-based addictions.

I've been treating Internet-related addiction problems since the mid-1990s, and after treating hundreds of patients, conducting research, publishing, and teaching behavioral addiction courses at a medical school, I've concluded that this addiction is treatable. However, unlike most other addictions (with perhaps the exception of a food/eating addiction), the trigger to use the Internet and your screens is literally ever-present in your daily life.

Also, unlike other behavioral addictions, such as gambling or substance-based addictions, the Internet *must be used in your daily life.* The challenge with Internet and technology addiction is in finding a way to limit the most problematic content, while still accessing necessary screen use. The following sections cover a wide range of alternative treatment options that you may consider.

Any treatment program or clinician should manage Internet and technology addiction as a primary diagnosis, while addressing other co-occurring problems as well. I discuss issues that co-occur with Internet and technology addiction later in this chapter.

Undergoing outpatient counseling and therapy

This is perhaps the most common approach used in treating Internet and technology addiction, and if appropriate, it can be quite effective. During the COVID pandemic, many of the outpatient treatments (and intensive outpatient programs) were held via telehealth video formats (which can be feel less effective but are way better than nothing and can still be helpful). Outpatient treatment usually involves one to three sessions per week, and we find that multiple sessions are needed in the earlier stages of treating Internet and technology addiction on an outpatient basis. In these earlier stages of treatment, more intensity and support are typically necessary to help establish the pattern disruption and cognitive restructuring that is needed.

There is a difference between managing the treatment of an Internet, video game, or technology addiction and treating other co-occurring mental health issues. Be sure to choose a doctor or therapist who has experience in treating Internet addiction *and* related disorders, and not simply the co-occurring symptoms of depression and anxiety, ADHD, Autism Spectrum Disorder (ASD), or other psychiatric issues. Many patients I've seen have received significant outpatient treatment focusing on *only* their psychiatric concerns, which did little to address the Internet and technology addiction. In my experience, focusing solely on mental health issues without addressing the primary addiction will prove unsuccessful; this is not to say that the other issues should not be addressed, but that it must be done as a comprehensive package of care to ensure the most positive outcome.

It is my strong recommendation that if outpatient treatment is to be utilized, it should be done in conjunction with an experienced IT professional, who can work directly with the doctor of therapist to help manage access if needed.

Considering intensive outpatient and partial hospital programs

We have been running a drive-in and fly-in intensive outpatient program (IOP) for the last 12 years at our Center for Internet and Technology Addiction. Our intensive outpatient program consists of a one-to-four-week clinic where the patient spends a large part of every day on-site (sleeping at a local hotel with a family member) and engaging in a variety of individualized therapies. This is not a group-based program. Typically, this is done over a two-week period, but it can be structured for longer or shorter periods as I indicate earlier in this chapter.

The more typical IOP or partial hospital program (PHP) runs for 30 to 90 days and involves going to a hospital or clinic on a two- to five-day-a-week basis for several hours a day to participate in a variety of group and individual therapies. More traditional IOPs or PHPs utilize a lot of group therapy and can be quite effective. Medications are also often evaluated and prescribed as needed. (I discuss medications and group therapy earlier in this chapter.)

REMEMBER

The challenge of most IOP-type programs is that you are still sleeping at home (in our program you are temporarily at a hotel) and in the same home environment that you typically stay in, so procedures need to be established for limited or modified access to screens — especially during the treatment period, and likely afterwards as well. This is where a knowledgeable IT person should be involved in the treatment planning and integrated with the doctor or therapist who is leading the treatment.

Looking at residential treatment centers

Residential treatment centers are very useful when treating Internet addiction and screen use disorders because they allow for removal of the temptation of screen access; however, they are temporary, and great care must be taken, and planning made, to help the patient develop a good discharge plan, which may also include installing filters, monitors, and blocks on the addict's devices at home or at school, once they are discharged. Even if you or your loved one is there for a few months, that time goes very fast, and one must almost start planning for discharge shortly after admission.

REMEMBER

There aren't very many fully dedicated residential programs for Internet and technology addiction. Last time I counted, there were fewer than six in the United States, and only a few that treated this issue exclusively. Care should be taken to avoid programs that do not have an expert in this field as a clinical or medical director or as a consultant; some residential and rehab centers will simply add to their marketing materials and websites that they treat this issue, but they really have little in the way of a specialized program or services. You want a program where the staff has some training and experience in treating these issues, and where there is an identified focus on Internet use disorders and video game addiction.

Conducting a Tech Sabbath

What is a Tech Sabbath? I really like this concept because it captures the idea of consciously taking a break from screens and technology. The idea is to create a *digital diet* whereby you elect to limit or reduce your use of devices (or specific apps) for a period of time; you will also want to avoid addictive websites or apps — such as YouTube, TikTok, Instagram, and so on — or whatever content areas are problematic to you. Choose to live certain hours, or parts of a day or week, with less (or even no) screen tech time. The idea is to have a set period where you're relatively tech free. These are partial ways to reduce some of your dependence on technology, but they are sometimes harder to accomplish when you're dealing with an active addiction. Largely, the technique is a reminder to your nervous system, and to *you,* that life can be okay (or better) with less tech for brief periods of time.

REMEMBER

The idea is to remind your nervous system that you can live with less tech, and *slower tech days* are an invitation to rekindle other methods of self-care, social interaction, creativity, and rest.

The interesting thing about Tech Sabbaths is that most people who use them report very powerful results. Although they are initially worried and anxious about turning off their phone, revoking an app or program, or turning off a notification, this generally proves to be less of a problem than they imagined. Initially, even taking a walk without their smartphone can be scary and create some discomfort, but after they have done so, I've *never* heard a negative result. Almost all people who reduced their tech and screen use feel better afterwards.

Remember the last time you got gas at the pump (assuming you don't have an electric car yet)? How long did it take? Perhaps three or four minutes, tops? Now some gas stations have decided you cannot possibly be bored that long and have installed TV screens broadcasting news, information, and ads on the stimulating Gas Station TV Network (GSTV). This is boredom intolerance at its best (or its worst), and it's another example of the spread of the erroneous belief that we cannot be still with ourselves for even a minute or two (especially if some advertising can get thrown in, too).

Examining weekend and extended detox

Weekend detox is an interesting concept. The idea that you structure short (but regular) detox periods, where you have little or no screen access for a brief time, can be effective. Sometimes it's done without any screen access at all. Typically, these detoxes are relatively short — perhaps a day or two.

Longer detoxes may require limited access for essential use of the Internet for banking, school, work, or email. This is tricky, however, because unless the other problematic content (such as video games, pornography, social media, and YouTube) is blocked, the addict may be strongly tempted to toggle to the other content areas.

REMEMBER

Ease of access is always a potential trigger for relapse.

Knowing about digital detox

A digital detox can be scheduled before or during treatment. As I mention in Chapter 12, a digital detox can be done simultaneously with the beginning of treatment, and it can be a full or partial detox. Some access to screens is typically needed, so it's advisable to arrange for an experienced IT professional to help set up specific app and website blocks and total screen time limits. Again, please do not attempt to do this without a doctor or therapist to help coordinate it all; these situations can be very emotional and taxing, and IT solutions in the absence of a therapeutic treatment plan are typically ineffective.

REMEMBER

Make sure that the IT professional and the doctor or therapist conducting the treatment are collaborating. I also strongly recommend that you consider using a trained IT person, as opposed to a knowledgeable parent or family member, in order to avoid problems that result from complex family dynamics. Some parents and family members elect to install their own blocks, filters, and limits; if you do, make sure the software you use is flexible and hack-proof enough to be reasonably reliable. Never use obvious passwords that the addict may know or guess.

Exploring wilderness retreats

Wilderness retreats are a growing area of personal growth and behavioral health support programs. They are not strictly a form of treatment. Rather, they are a type of program involving outdoor and adventure-based education where self-reliance, team building, stress management, and social skills development can all be addressed while engaging in outdoor physical and survival activities. These programs can be very challenging and are generally rough (meaning they lack creature comforts), which may be fine. Because they are mostly outdoors, they do offer a significant buffer from screens and technology.

WARNING

The problem I've found with wilderness programs is that they are not particularly therapeutic; rather, they are based on the idea that removing the addict from their screens helps. It does help, but unless there is some treatment or intervention regarding home or school use of screens, the previous use and addictive pattern returns shortly after they complete the wilderness program.

Working with a coach

What is coaching? Well, it's exactly what it sounds like. A coach is a person who acts as a consultant or guide to help facilitate dealing with a particular problem or circumstance. Typically, a coach has expertise and experience in dealing with a certain problem or issue, and sometimes works with other behavioral health professionals in dealing with medical, psychiatric, or addiction issues. Coaches can be a great support for parents who need help in setting and managing rules and limits regarding home screen use.

WARNING

Coaching is somewhat controversial. You do not need a medical or graduate degree to be a coach, and to my knowledge no state license or certification is needed to be a coach. Several organizations privately certify coaches, but it is pretty much up to the consumer to verify the level of expertise that a coach might provide. Many coaches do have some academic or clinical training as well.

Exploring EMDR

Our clinic has been using the neurophysiological *eye movement desensitization reprocessing* (EMDR) procedure for the last 20 years, with good results. We typically use it as part of a comprehensive treatment plan for Internet and technology addiction to help reduce urges and cravings in order to reduce potential relapse, and to address psychological trauma and/or anxiety (including generalized and social anxiety).

EMDR can be a very powerful and effective neurophysiologic technique to help neutralize and desensitize traumatic memories and emotions, reduce anxieties of various types, and decrease the intensity of addiction urges, cravings, and triggers. EMDR is approved as an effective first-line treatment for trauma and post-traumatic stress syndrome by the American Psychiatric Association.

The first part of EMDR involves taking a detailed history of the problem (which is usually already done if treatment has commenced). Part of the EMDR procedure involves inducing rapid eye movement (using bilateral light stimulation), which triggers eye movement patterns like what occurs during REM (rapid eye movement) sleep. Sometimes bilateral auditory or tactile stimulation can be used in addition to or instead of the eye movements.

The EMDR processing continues during the treatment process, which is typically conducted in 90-minute sessions until the troublesome emotions, urges, or triggers are addressed. It is believed EMDR helps support an accelerated processing of memory and emotions, which can speed up the psychological healing process. Neurobiologically, EMDR appears to reverse reciprocal suppression of the anterior cingulate cortex, which is an area of the brain implicated in impulse control, emotion, and decision-making. It's interesting to note that I've seen some remarkable results with EMDR reducing impulsive and compulsive screen use, which may be related to the impact of the procedure on the anterior cingulate cortex.

Investigating interventions

An intervention is a method where an addict is confronted by friends, family, and loved ones regarding their addictive behavior. It is usually done in a group setting and enables family and friends to express their concerns and fears about their loved one. Interventions are usually run by a professional who has experience with organizing interventions for various addictions. To my knowledge, few interventionists specialize only in Internet and technology addiction; rather, they may more likely deal with a variety of addictions and behavioral health issues. Interventionists will sometimes arrange for transportation to a residential treatment facility, assuming the addict is willing to go.

Make sure you ask detailed questions of how the interventionist works and their overall approach. No state licensing is required to be an interventionist, although some of them have a mental health or addictions clinical background or have been in some type of recovery themselves. Sometimes interventionists have relationships with several facilities, doctors, and therapists, and they may also serve as a resource for services.

REMEMBER

Interventions are not always successful, and they may not convince the addict that they do, in fact, have a problem, but they are a tool that can be used when the addict is in denial.

Treating Co-occurring Psychological Issues Connected to Internet Addiction

The idea that underlying psychiatric issues might be contributing to an Internet and technology addiction is complex; it is complicated because *contributory* factors can become *resultant* factors, and this creates an interactive and recursive loop between cause and effect. It is better for the doctor or therapist to do an assessment to evaluate the current symptoms and to create a treatment plan to address the primary addiction, as well as the other psychiatric problems that may be occurring.

The point is that once the addiction has started, it is difficult to discern between etiological and secondary symptoms; in my opinion, they should all be treated as co-occurring symptoms and managed as part of the overall treatment plan. The following sections cover some typical co-occurring issues with Internet and technology addiction.

Understanding anxiety

Anxiety is a set of symptoms that often involve physiological arousal responses, fears, and associated negative thoughts. Anxiety is often linked to concern or fear about the future or some future event. In other words, anxiety about what might happen or could happen is not always rational or logical, but it may be based on a kernel of truth and become exaggerated based on irrational fears. Anxieties are often specific to certain fears such as social anxiety or performance anxiety, and anxiety disorders may also be generalized, exhibiting a nonspecific or free-floating sense of dread or fear. Sometimes anxieties can develop into a phobia involving a social situation or other circumstance, such fear of flying, insects, or other focused situations or circumstances.

Anxiety disorders come in a variety of types, and although they all have unique qualities, they typically all share a fear-based physiological reaction to a situation or circumstance and involve a general emotional response. Social anxiety is perhaps the most common form of anxiety that we see with Internet and technology addiction patients; it is not uncommon for upwards of 50 percent or more of the cases that we see for consultation at our clinic to have some history of social anxiety.

REMEMBER

When social anxiety becomes extreme, it can sometimes escalate to a level of avoidance of social interactions. Typically, the social avoidance and anxiety that we see with these individuals does not apply to social interactions that occur via the Internet; they find their screen relationships and interactions through social media, video game platforms (such as Discord or Mumble), and other online interactions to feel safer and more comfortable. The reason such onscreen activities are tolerable may be because they pose a less direct threat than the emotional intensity that often accompanies real-time, social interaction.

Digging into depression

Depression and depressive disorders are more complex in the sense that very often, individuals that we see for Internet and technology addiction treatment or consultation have some degree of depressive symptoms upon initial presentation in our clinic. This may be partially due to reward deficiency and the reduction of available dopamine receptors (thus less available dopamine) for real-time living. These individuals are seemingly unable to absorb pleasure from interactions that do not provide the dopaminergic power that they get from online screen behavior. Often it takes a period of detox for neural pathways to re-regulate and to see what their baseline level of mood is. Sometimes these individuals have a prior history of depression, although it is very hard to ascertain what their baseline level of mood is — very often, the patients we see have been using screens intensively since they were 5 years old.

REMEMBER

Time is needed to see what the underlying level of mood is for these individuals, and whether depression is the *result* or *cause* of some of their addictive screen use. Often, I find that individuals who end up at our clinic have had depression listed as being their primary diagnosis, and that their video gaming or excessive Internet use was seen as secondary to their depression. I'm not sure that this is always the case; I think that just as often, if not perhaps more often, these individuals may be depressed *because of* their excessive screen access.

The only way to know for sure is to spend adequate time evaluating the patient and taking a good medical and psychiatric history. The problem is complicated because when you isolate yourself, you limit your stimulation and emotional regulation to one mood state (on screens), and so it is difficult to know whether they are experiencing normal mood fluctuations or not.

REMEMBER

Obviously, if there is an active depression, this needs to be addressed concurrently along with the addiction. It is my experience that treating the addiction without addressing the depression, or addressing the depression without addressing the addiction, will likely fail and ultimately lead to further relapse. It is always important that both be treated simultaneously.

REMEMBER

I want to talk a bit about depression, which might include suicidal behavior or potential for self-harm. One of the hallmarks and risks of depression is the possibility of suicidal thoughts, gestures, or attempts; this needs to be taken *very seriously,* and a thorough evaluation needs to be made around safety and risk for self-harm. This is part of a standard psychiatric evaluation and is typically done by most mental health and addiction clinicians upon the initial consultation or at reevaluation.

One of the things I have observed with heavy Internet and video game users is that when we discuss the idea of having limited use, digital detox, or potential treatment, there can be a *direct* or *veiled* threat that they will do harm to themselves if they are not allowed to continue with their typical screen habits. This is very tricky and requires very careful evaluation because, above all, *safety* is the first concern, and we want to make sure that patients are safe and protected from doing harm to themselves. The details involved in a safety and risk evaluation are beyond the scope of this book, but any competent mental health and addiction clinician who will be evaluating you or your loved one can attempt to ascertain potential risk.

REMEMBER

Lastly, if you are ever seriously concerned about your or your loved one's safety or risk regarding suicide or self-harm, never take a chance; call 911, your local crisis team, or the police, or, if appropriate, transport the person to the nearest emergency room. It is *impossible* to accurately predict suicidal or self-harming behavior, so a *conservative* approach is always indicated; their Internet addiction can wait and be addressed after they are evaluated and deemed safe.

Talking about obsessive-compulsive and impulse control disorders

Impulse control disorder and obsessive-compulsive disorder are conditions that I tend not to see as often as other types of psychiatric problems such as anxiety, depression, and ADHD. I've seen several cases that have some features of obsessive-compulsive disorder, but these symptoms typically haven't revolved around the Internet and screens.

Impulse control disorder, however, is a general category that could include the impulse to engage in various online screen behaviors to manage internal emotional states. However, with Internet and technology addiction, we don't only see impulses to act, but rather a repeated pattern of compulsive and persistent behavior. Very often, we will diagnose an individual as having an impulse control disorder to capture the intensity of their drive to use screens, but this doesn't fully capture the all-encompassing quality that we see with an addiction.

Studying ADHD

REMEMBER

Attention deficit hyperactivity disorder (ADHD) involves a variety of symptoms, and in my experience, 85 percent of all patients I've seen in the last 25 years for screen-related issues carry a diagnosis of ADHD either before they come to see us or upon diagnosis at our clinic. In general, the frequency and co-occurrence of ADHD and Internet and technology addiction are so high that they undoubtedly have a shared and interactive relationship on a neurobiological level. Whether ADHD adds to the propensity for developing a screen-based addiction or vice versa is unclear.

There are several different types of ADHD, the three most common being ADHD hyperactive type, ADHD inattentive type, and ADHD combined type. The combined type would involve both excessive motor movement and difficulty with attention and concentration. The most notable feature of ADHD is a difficulty in being able to attend, concentrate, and focus. Extraneous or distracting stimuli are very hard to filter out. The ability to deal with routine or monotonous activity is difficult, and there is a high threshold for stimulation; the patient needs information to be of a *highly stimulating nature* to become strong enough for them to be able to attend to it.

In ADHD, there seem to be deficiencies in dopamine in the orbital prefrontal area of the brain. Internet and video game addiction is likely prevalent among the ADHD population because of the higher stimulatory threshold seen in ADHD patients, and the high stimulatory potency of the Internet and video games. If you've never seen a video game on a computer screen, do yourself a favor and look at one, because it is very understandable why an individual who has a hard time attending to routine stimuli would like a video game; even looking at YouTube videos or just surfing the Internet provides the dynamic novelty and stimulation that ADHD patients love.

The other thing that is notable about this population is that they love novelty and variability, and the Internet is full of novel, dynamic, and variable content — it has endless options, with unending opportunities. With the Internet's ability to capture attention and provide stimulating content, it's a perfect self-medicating drug for those who have the attentional issues seen with ADHD.

REMEMBER

Lastly, there is some preliminary evidence to suggest that not only does ADHD increase risk for Internet addiction, but the inverse may also be true — excessive Internet and screen use may decrease attentional capacity and, in a sense, cause or exacerbate ADHD symptoms.

Seeing the connection with autism spectrum disorder

Another condition often seen in our patients for Internet addiction evaluation is a history of Autism Spectrum Disorder (ASD). Most patients we have seen with this diagnosis or who have been diagnosed by us during our evaluation typically fall into the mild level. There also seems to be an over-representation of ASD diagnosed (or undiagnosed, but symptomatic) with Internet and video game addiction; this may be due to some of the neuroatypical processing that we see with ASD individuals.

I am beginning to look at some of the neurobiology that is associated with ASD and the similarity with some of the deficiencies that we see with ADHD (see the previous section). The properties found in Internet and screen communication may be a perfect fit for some of the limitations and deficiencies found with ASD patients — if you have communication deficiencies, as well as difficulty managing social cues, then the Internet, and the limited frame within which the Internet operates, is a perfect circumscribed modality for people with ASD. The problem is that the Internet is so addictive, which may block the patient's desire and ability to communicate using other real-time methods.

Noting other addictive behaviors and substances

In general, I have not found the population of Internet, video game, and technology addiction patients to be at a substantially higher risk of developing other substance use and addictive problems; there are some exceptions, however, and I've seen instances where other behavioral addictions, such as gambling, preexisted their Internet addiction problem. I've seen only occasional cases (less than 10 percent) with substance use disorders, most typically marijuana. I cannot recall an instance of excessive alcohol consumption by patients who I have seen at the clinic, although we certainly include all substance-based and behavioral addictions as part of our routine evaluation.

In general, it has been my experience that the overall risk level for alcohol abuse, substance abuse, and other addictive behaviors seems average in this population. Although this is not an evidence-based conclusion, it is based on my clinical experience over the last 25 years. Perhaps heavy substance users typically go to other facilities where it may be revealed upon their evaluation that they are also addicted to screen use, or once they get sober, they use the Internet and screens as an alternate source of dopamine stimulation.

One final note: It is possible that the dopaminergic innervation that is being provided through excess screen use obviates the need for obtaining intoxication through other substances and behaviors. In other words, the medication that is being provided by screens may be enough. Not that this makes addictive screen behavior a good thing, but it does seem to provide some of the mood-altering properties that are often sought out from addictive behaviors and substances. (See Chapter 3 for more information.)

4

Living a Balanced Life with Internet Use in Its Proper Place

Find some suggestions for managing screen overuse as well as treatment options for Internet and video game addiction. Rediscovering real-time living opportunities and learning to manage boredom are good beginnings for finding life beyond mindless screen use.

Get suggestions for how to parent in the screen age, and for ways to deal with having a screen-addicted child or loved one. One of the biggest challenges parents face is how to manage their child's screen time and manage access to their Internet devices. This is particularly problematic when it comes to addictive content such as video games, pornography, social media, and binge-streaming TV.

Understand some of the future changes and developments that are likely to occur in the world of Internet screen devices. It seems clear that Internet technologies are not going anywhere and that people will need to learn to manage their use of these devices in order to maintain a healthy tech-life balance.

IN THIS CHAPTER

» Knowing why you may need to change your tech use

» Becoming honest with yourself about your screen time

» Getting a handle on your goals and values

» Developing real-life activities to do in place of tech

» Beginning and keeping good screen use habits

Chapter **14**

Solutions for Real-Time Living

This chapter focuses on what I like to call *Tech Rx*. What is Tech Rx, you ask? It's a set of tools and attitudes to help manage the excess screen use that we *all* find ourselves involved in these days. Even if you aren't addicted to Internet technology, based on current-use statistics, an antidote to the excesses of being screen-bound is clearly needed. Just look around — a lot of us look like *tech zombies.* I'm still surprised at how many of us have our faces buried in a screen for endless hours a day; people spend a lot more time on their screens than they realize, and they clearly need to use tools and solutions to manage their use so that it does not manage them. I'm talking about regaining our freedom and independence from the silent control that our devices have on us.

If you continue to allow your screens to take up all your time, you will continue to see a growing imbalance in yourself and your children. There is a *limited* amount of time (and attention) in each day, and the ever-increasing amount of time that people are spending on their devices must come from somewhere. Often, it comes from self-care, other forms of relaxation, exercise, real-time relationships, sleep, and productivity.

All this screen tech can come at the price of real-time social interaction. People now have blazingly fast Internet and portable high-speed access for their screens (and it's getting faster all the time), but they have *fewer* intimate friends and social relationships than ever before. It's far easier to turn on TikTok for an hour or two than to make plans to meet a friend for coffee. You may not realize it, but your technology has slowly and silently changed you. It has gradually usurped some of your essential humanity and replaced it with *infotainment.* It seemed innocuous at first, but how innocuous can it be when your technology is now shaping your daily schedule, instead of the other way around? How free are you if you cannot leave your phone for a few short minutes or if you find yourself looking at one of your Internet-connected screens hours on end?

REMEMBER

If you don't manage your technology and screen use, it will *always* manage you.

Reminding Yourself of the Importance of Changing Your Tech Habits

Many solutions exist to *limit* and *manage* your screen time. These include various software programs and apps, and most smartphone operating systems or service providers have screen-time apps and blocking capacity built right into phones. But I've found that very few people really look at these available forms of feedback, and just like you need feedback on your weight or what you eat, you need the ability to see how much time you spend on your devices in order to make any changes. Remember, most of us cannot keep track of how much time we spend on our screens. Sounds simple, right?

Wrong. Again, most people simply do not look, or if they do look, it is infrequent. And then, despite getting upset in the moment at seeing that they have spent *seven hours* on their smartphone that day, they soon forget and end up doing the same thing the next day.

REMEMBER

You can no longer afford to see your digital screen devices, and the Internet in general, as harmless; these are potent and powerful *digital drugs* that need to be respected. The following sections remind you of the power of screens and why it's wise to check in and change your habits.

Getting an accurate picture of your tech time

You need three things to really change your screen patterns and habits. Whether you are simply overusing or are addicted, the process of change is the same:

1. You need to *accurately see* what you are doing by taking a *fearless* look at how much time you are wasting on your screens.

2. You must decide that you want to change *what* you're doing.

3. You must learn to substitute *real-time living* and *being* in place of using your screens.

Painful information is just that: It is painful, and people do not like pain. They tend to avoid painful things — even if it is in their best interest not to. The minute they see a painful truth, people have *cognitive defense mechanisms* that rush in to *rationalize* their reality and to minimize any psychological pain or discomfort.

You will probably have rationalizations and reasons for being on your devices, but I'm not talking about necessary use for work or school. Most of the time that you spend (waste) on your devices is probably *not* for productive use. It is all the social media time; the deep dives into Wikipedia, Reddit, or even Google searches; the auto-advancing YouTube or Netflix playlists; or getting sucked down the rabbit hole of gamified news apps, with endless click-bait and related links, that are *designed* to keep you in the cyberspace maze for the next ten years. All content producers and service providers *want you* on your screen — all the time!

All apps and websites are designed for one purpose and one purpose only: Regardless of the content they offer, they are using the addictive Internet medium, with as many tricks and detours as needed, to keep your attention. Social media may have been invented for social purposes, but it has morphed way beyond its original purpose. The *Like* button on Facebook and the *Streak* feature on Snapchat are designed to do one thing: to make sure you stay on *longer* and *more often*.

Sure, these apps can be fun. Sure, they are pastimes. But are they valuable? Are they socially nutritional in any way? Do they make your life any better (or happier)? Perhaps they do in some ways, but I contend that they are so powerfully addictive, and they are such time-sucks, that *you really cannot objectively perceive how much is too much.* You lose the ability to tell when you're reading the news versus when you're being sold lawn furniture.

I don't know about you, but it angers me when I'm being manipulated. I don't like it when the million-year-old reward center in my brain is being tweaked to market to me — and it's done in the name of news, or social relationships, or something else that I may think is important. I know this is how advertising works, but this is a much more efficient and insidious pathway into your nervous system. This is the essence of manipulation, and it is potent and purposeful. Unless you see it for what it is, you'll be wandering around the slot machine called TikTok for hours, wondering what the heck you are doing there.

Digging into the "screen drunk" phenomenon

We all laugh about how addictive and compelling this technology is, but is it funny? Is it funny that you can waste hours of sleep time looking at AI-based algorithms designed to feed you what you have shown the AI system you want? Sure, that neat TikTok or YouTube algorithm learns what you like and then gives it to you (even Amazon and Google do this), but it is not out of love or concern for you — it is *purposefully commercial,* and you had better remember that. You *can* begin to consider taking small steps to rein in who or what you give your time and attention to.

It is amazing that a person can be staring at a screen for 4, 8, 12, or more hours a day (not for work or school) and be unaware of how much time they spend. Few people can accurately see how much time they are spending on their screens. This phenomenon is called *dissociation* or *time distortion* and is discussed in Chapters 1 and 2 — and it is the way your brain distorts the passage of time when in a dopamine-stimulated state.

This is very similar to the experience people have with drunk driving. To study why people would drink and drive (this could also be applied to distracted smartphone driving), researchers gave subjects progressively increasing doses of alcohol and had them rate their level of intoxication and impairment. They were also given visual-spatial-motor tasks, and their performance and accuracy were measured. Guess what happened? They had no idea how drunk or impaired they were, and they had no idea how badly they performed on the neuropsychological tests they were given.

WARNING

The point is that people are very bad judges of their attention and functioning when intoxicated in any way, and I contend that smartphones and other Internet screen devices leave people significantly *intexticated,* such that they are unaware of how and what they are doing. It is a perfect captive opportunity to sell you things while you are in an altered state of consciousness, when you are, for all intents and purposes, *screen drunk.*

As noted in Chapter 10, the research that I conducted in 2014 and 2015 with AT&T titled *It Can Wait* showed that people had a very similar belief about distracted smartphone driving, but with a little twist: They knew it was dangerous to use their devices while driving, but over 75 percent of them did it anyway! Even the knowledge of impairment does not always change addictive behavior.

TECHNICAL STUFF

If you're a little older, you may recall ads for cigarettes on TV; this is no longer done, and in general, people have decided that it isn't okay to advertise addictive and mood-altering substances and behaviors. Many of these advertisements have been eliminated and others are more regulated, but video games and all things screen-based are still fair game. How many ads do you see weekly for cellular service and smartphones? At the time of this writing, a bill, the Children and Media Research Advancement Act (which I endorsed), is pending in Congress and is intended to support research on this issue. When it comes to children and screens, I suspect many changes will occur over the coming decade in relation to advertising and the addictive properties of screen technologies.

Balancing Your Values with Your Tech Use

WARNING

Some people may ask, "What's the harm? It's only a screen." However, that *is* the problem. People see their screens as sort of benign friends that entertain and distract them, but what they do not see is that their screens are accomplishing this by using a *velvet handcuff.* You are not as free as you think you are if you cannot leave your phone or tablet at home, or turn it off, or not use it for a day (if you doubt this, just try it); most people become anxious at even the thought of this, and they also become agitated if they think they left their phone somewhere. You might also find it hard to believe that there is an entire multibillion-dollar industry built around *getting* you on and *keeping* you on your screens. And that is the problem — people do not think about how much their devices control them.

I was recently thinking about how when I grew up, we had *one* landline phone in the house, then we got extension phones in other rooms that gave us freedom to move around the house. I still recall the excitement I felt as a teenager when I installed my own extension phone in my bedroom (yes, I installed it myself). Then came the wireless phone for home, which gave us untethered freedom. Fast-forward a few years to 2008, and enter the smartphone; suddenly it became the *main* (and only) number we used for all our calls, so we all ditched our home landlines. Now we carry our smartphones everywhere and around our homes so we never miss a text, call, update, or notification — we are essentially back to having *one* phone again. I think about this each time my cellphone rings or a text comes in and I left it in another part of the house; suddenly my cellphone no longer feels *all that convenient* as I begin to feel tethered to it, even if I don't want to be.

Do you value your time? Most people would like to say yes, of course they do. This may be true, but I suspect some of it is cognitive dissonance, where they feel they should value their time when, in fact, their behavior may not actually support that reality.

The funny thing when it comes to technology use is that people often decide their values based on how they spend their time, and not the other way around. This is backwards, as they end up justifying how much time they spend based on their addictive or excessive screen use — often missing the opportunity to create a tech-life plan based on their larger goals and values. If you end up *passively* thinking this way, your screen time use will continue to be based on an *automatic* and addictive behavior, and you'll remain caught in a virtual loop of endless overuse, with little to show for it.

Your tech use *needs* to be part of your daily life. It's impossible to live and work today without your screens being connected to the Internet. But I would argue that it's also hard to live in a balanced and productive manner without creating a *values-based approach* to your technology.

Follow these steps to help develop a values-based approach to technology:

1. **Look at your overall list of values and make a list of categories.**

   These might include work, school, exercise, self-care and personal hygiene, social relationships and family time, marriage and intimate relationships, creative tasks, entertainment, religious or spiritual practices, food shopping and preparation, and lastly, sleep. Add any others that are specific to your life situation.

2. **Look at the approximate time commitments these activities require in your day and week.**

   I think you'll find that the three to seven hours a day (or more) during which most of you have been using your screen devices (and that's typically just your phone) may be compressing or eliminating your ability to attend to your values and goals for daily living. Something must give, and it's *not* your screen. The problem is the way people are using technology — they treat it like a *benign pastime* with little awareness of their heavy use, and they put it at the *top* of their daily activity instead the bottom.

Notifications on your smartphone, tablet, and computer serve one function only: to get you to look at that screen, and hopefully get you stuck looking for a while. Ever notice that once you look at your screen, you seem to keep looking, and often longer than you expected? That is the neurological power of the Internet, and as I discuss in Chapter 2, this keeps you coming back for more. A notification gets that engine going and causes you to look, and then you are off and running. The only good solution is to turn notifications off (or significantly limit them) and begin to ignore the powerful FOMO (fear of missing out) or NOMOPHOBIA (fear of no mobile phone). Trust me, you will miss very little of true importance without notifications, and you will have much to gain with the time and attention you will be reclaiming.

Getting Real with the Real-Time 100

In Chapter 12, I introduce the *Real-Time 100 Living Plan.* This is a simple tool you can use to write down a list of behaviors you can engage in that don't involve screens. The idea is to remind yourself that there are *many* things you can do that may be more consistent with your overall life-values and that do not involve mindlessly staring at a screen.

As an addiction psychologist, one of the most common complaints I hear from my patients dealing with life stress is that they wished they had more time to do the things they always wanted to do but could never find the time. Sometimes I ask them to take out their phone and to scroll to the screen-time app or other app that records their overall screen use. Usually, I ask them to tell me how many hours a day they are spending, and then to look at the weekly figures. Initially, they are shocked, and then embarrassed, because invariably they are spending between 15 and 40 hours or more a week just on their smartphone and apps, and that does not include other devices.

REMEMBER

Think about it. Twenty hours a week is about 3 hours a day. You have 168 hours a week. That is it; no one gets more. If you sleep an average of 8 hours a day, that leaves 112 waking hours a week, and when you spend just 3 hours a day online (21 hours a week), that is about 20 percent of your weekly waking time on a screen. Three hours is on the low side. Let's say you're closer to 40 hours a week or about 6 hours a day, which means that nearly 40 percent of your available waking time is spent on a screen — how can that allow for all the other valued aspects of your life?

Given that you're awake about 112 hours a week, in a year, at 3 hours of use a day, assuming you live until age 80, you'll spend about 87,000 hours on your screens; at 6 hours, that becomes about 175,000 hours. Those numbers are slightly inflated because you aren't likely to have used your devices as a very young child, but you get the basic point here (you can reduce the numbers by 8 percent, assuming you start using screens at age 10). Of the approximately 467,000 waking hours (assuming 8 hours sleep a night) that you have in an 80-year life span, at just 6 hours a day, you'll spend about *30 percent* of that available waking time on your screens. Do the math. The question you should ask yourself is whether that is what you want to be doing with your limited time on earth.

Instead, you can create a lower-tech living plan — not simply a list, but a plan that includes things you *like*, things you *love* to do, and the people you like to *do* them with. Time is limited, and the Internet can eat what time you have without your even realizing it.

The way to live with less tech is to *practice living life* without it, or at least without as much of it. Start small, and try these tips:

>> Practice taking a walk, walking your dog, being in a waiting room, waiting in a line, sitting and just thinking, or having a meal or a cup of coffee *without* your phone. Try to *pair fewer activities* with your phone, tablet, or laptop, and consciously look for opportunities to do it differently. This will feel odd at first, but remember when smartphones first came out; initially it felt weird to have them out in public.

>> When you sit down at night to watch a movie or TV show, resist the temptation to look up every actor or obscure fact, order an Amazon item, or do a social media check-in on your phone, tablet, or laptop. Focus on *one* task or activity at a time.

>> When you go to bed, leave your phone charging in the other room. Buy an alarm clock. You'll sleep better.

>> Consider leaving your phone in the car when you meet your family or a friend for dinner. Consider never having your phone out when eating or socializing; this conveys a mixed message to whomever you are with — that you are physically there, but not fully present.

>> Turn your smartphone screen to black and white (gray scale); this makes it less appealing and will quickly cut down your phone use. It's an easy setting change (yes, you may need to look online to see how to do it for your model).

I know these are hard things to do; I've done them, and they feel strange at first. However, if you don't practice creating some space and some distance from your technology, it will continue to envelop your life and eat up much of your time. It's not all or none — any change is a positive change.

Building and Maintaining Good Tech Habits

Technology is like any other aspect of living; it requires some positive habits and skills to master its use. You must ask yourself what you want to do in your life and then look at whether your Internet use will get you *closer* or *further* from it. If technology use only improves the quality of your life, then that is a good thing, but it clearly doesn't always work that way. It certainly can be entertaining; it can forestall boredom; it can educate you and help you be productive. But it can also take you away from real-time relationships, reduce intimacy with your loved ones, lower your self-esteem and empathy, increase depression, impact your sleep, and make you more sedentary.

Your screens should be *tools* that you use to improve your life; just like a glass of wine can improve a meal, screens can enhance your *digital diet* if they're kept to a portion of your daily living. The following sections provide some insight into and guidance for better tech habits.

Searching for the humanity in technology

I remember the early days of social media when it was hailed as a great advance in human connection. I think we have come far afield from those early days, where people assumed that social media would help them stay connected to all those people they love and want to share time with. This is a great idea, and certainly does occur in many ways, but it just didn't work out quite as originally planned. Nowadays it's more about *social commerce* and *entertainment* than social connection.

When I first got Facebook, I recall being initially amazed at being able to find old friends and to keep family updated with photos and personal news; it was fun, and during the COVID pandemic, people's screens proved invaluable for work and school and for sharing time with friends and family. The COVID pandemic allowed us perhaps some of the greatest achievements yet from Internet and screen technology, but most of the time spent on screens is generally *not* for all these great things. It is to play video or smartphone games, look at YouTube, stare at TikTok, and get lost for hours on Wikipedia, Instagram, Facebook, Reddit, Twitter, Amazon, and watching pet videos. People waste hours picking out the news from in between the ads and click-bait. They mindlessly scroll, surf, and look; what they are looking for is to feel something — to feel good — and in the moment, their screens keep them just distracted enough (and perhaps numb enough) to avoid the urge for something more, something greater and more sustainable. Screens can numb you, similar to people who sit in front of a slot machine all day and night at a casino.

All the time you spend interacting on your screens not only wastes a lot of time, but it also doesn't reduce the need you have for real-time social interaction. The problem is that you use *so much* time on your devices that you have little time left for more nutritional connections. You also have the illusion that what you're doing is socially connecting, yet it doesn't provide the same benefits. This is like when you are hungry and eat empty fast-food calories; you're temporarily satisfied, but you still haven't consumed the nutrition you need. Much of your screen use fills you up (uses your time), but unless you question what you're doing and how it fits in with your life, you'll be left wasting a whole lot of your valuable time.

REMEMBER

I don't believe you'll ever find what you're looking for on the Internet or your screens. I've looked, and although it can be cool, fun, and interesting (as well as useful and productive), in my estimation, it is *not* life-enhancing in the way a good friend or family is, the way a good book or movie can be, or how a walk, conversation, or just sitting and watching a sunset can be. That is not to say that for many people, their ability to use their screens can be freeing and amazing; however, I do not believe that it gives many of us enough of the nutritious stuff needed to make us feel sustainably connected in the world.

Being mindful with your tech use

What is mindfulness in tech use? It's the same as mindfulness in everything you do. What I'm suggesting is that you bring *conscious choice* to your technology. Not autopilot, where you mindlessly use your screens and then find that you have spent half your waking hours that day fooling around on your phone or tablet, or spending the better part of a day on a video game or watching YouTube videos. It is choosing and being conscious of your screen-time choices.

REMEMBER

Mindfulness in tech use creates a sustainable pattern of screen use where you can count *on* your technology when you need it and count it *out* when you do not. What you think constitutes need is often simply a habit that you have on autopilot. Sustainable tech use is the ability to use your technology responsibly and to not overuse or abuse it.

In study after study, teens and young adults who voluntarily reduce their Internet and screen use report feeling *happier* and more *satisfied*. People who gave up their social media feel better for having done so; they report feeling like they got off an endless hamster wheel. At first, they are all reluctant and even angry about the idea of using less Internet and screens, but invariably, they feel better for it after the fact. Try it and see whether you experience something similar. I've largely stopped using social media; occasionally I do use it (I went on for my birthday recently to read the good wishes and thank people), but generally, I try to avoid the sinkhole we call social media.

Keeping the Internet as a tool, not a destination

People seem to have forgotten what the Internet and their screens are supposed to be for. They are supposed to be tools, not lifestyles. A substantial amount of research shows that too many people see their technology not as a tool, but rather as a *friend* or a *place* to be. The problem is that if the Internet, with all its apps and websites, becomes a habitual place to be, *where* are we living our real lives?

Tools are great. Tools help you do a job, and the Internet and all your screen devices are amazing tools. But that is it. They are tools you can use to help make your life productive, happier, and healthier; however, spending countless hours on your screens for the better part your day does *not* help you be happier (or healthier for that matter). The more you separate from yourself and others (and tech does just that), the less happy, connected, and satisfied you are.

REMEMBER

If cyberspace is where you spend your time, then the Internet is no longer a means to simply get something or do something — it has become a *destination* in and of itself.

Living virtually is a convenient way to be part of things you enjoy without the complexity of real-time interactions. In many ways, this makes sense, and in small doses it offers a great advantage. The problem is that it loses the richness and nuance of the connection that occurs in face-to-face interaction. I'm not saying that virtual connection is better or worse, but rather, that they are very different. I believe that it is this difference that found Zoom learning during COVID not ultimately as successful as real-time learning. It has its place, but it cannot be used *in place of* real-time interaction.

During the COVID pandemic, my medical lectures and teaching had to move to a virtual or webinar format. It was great to be able to still teach and do what I love to do, and it was also *very convenient.* I didn't have to drive or fly anywhere, and it saved a lot of time and expense, but if I must be completely honest, it was not nearly as much fun. It's impossible to read the audience via a screen — to see the faces of people and to adjust my words or style in accordance with their verbal and nonverbal feedback. I couldn't tell whether they were impacted by what I was saying, because it felt like a one-way dialogue. That is not to say that video learning and remote business meetings are not going to continue to be a game changer, because I think they will. Telehealth and telemedicine platforms also allowed me to see patients, and this was a wonderful option that clearly has its place, but there is something flat about such screen interactions.

People need *social* interaction. I do not mean Facebook, Instagram, Snapchat, TikTok, and LinkedIn; rather, they need to have some visceral interaction with people. I think when all is said and done regarding Zoom, Skype, Microsoft Teams, Webex, FaceTime, and other video platforms, people will say they were a life saver (and they were), but that they miss working, learning, and relating in person.

REMEMBER

I'm not saying you must put down all your screens. But I am saying you need to put down your screens more than you have been doing.

Overcoming urges, cravings, and boredom

Of all the triggers and urges I see with excess screen time, boredom seems to be at the top of the list. When I say boredom, I'm not talking about having hours of excess time with nothing to do. What I mean is that we have become a *boredom-intolerant* world, with barely the ability to sit for a few seconds without reaching for a screen.

REMEMBER

To change boredom intolerance, you must first resist the urge to scratch the digital itch. The only way to do this is to become consciously aware that you're experiencing the urge, and to create enough of a *conscious pause* to hold off picking up a phone, laptop, or tablet. It isn't easy, but it can be done with some effort. Even holding off a bit can create enough space to allow you to make a different choice. It is *your* choice. Ease of access increases the addictive potential of any substance or behavior, so find a way to create some space between the urge and the screen. Consider the following ideas:

>> **Making it a little harder:** One way to create enough of a gap is to not have the phone so close to you all the time. I'm not saying to have it miles away, but rather to move it a few feet away or in another room. One thing I do is to put the phone in another room or in my backpack, or turn it off or silence it. Sometimes I'll choose to leave it in car so I won't be tempted to use it so easily. Make it just a little harder to get to. I promise it will be there when you get back. And part of learning to do this is to educate friends, family, and associates that you may not *always* be able to respond in three seconds. This *instant response promise* has become the dominant expectation of our digital age, where *fast* is good, but *immediate* is better. Think about it: In about 25 short years, we've changed the conventional way of living and working, and not entirely for the better.

The idea is to put in some natural space between you and your screen until you have a better ability to deal with *moments* of boredom. Over time, you'll get better at doing nothing for a moment or two, and then those moments will become longer. The more space you can create without jumping to your screen, the more opportunity you can have for real-time social connection, self-reflection, and creativity. The numbing ability of your screens thwarts your desire and ability to extend yourself and to reach toward bigger, and perhaps better, uses of your time.

>> **Getting angry:** When I look at my screen time and it's greater than I want, I think about all the other things I could have done with that time. I get angry at the ability of that screen to guide my life in directions that aren't in my best interest. Anger can be a motivator to action. Anger can remind you that your time is valuable, and that *you* should decide where to use it, not an corporate tech algorithm. Having some anger at how you're being digitally manipulated by the tech industry should give you some pause.

>> **Creating a call to action:** The problem with all the screens you consume is that they are so easy to use, and it is a *passive and effortless process* to boot. It takes less energy to pick up your device and mindlessly watch or scroll for a few hours than to turn it off, leave it in another room, or do something different. Part of the solution is to change what might be termed a *passive* (but potentially addictive) activity into an *active* choice that requires doing what is initially more effortful than being a passive recipient. Those choices must start first with becoming *conscious* of how and what you're doing online and then to decide whether you're satisfied with the resulting outcome. If addicted, the choice is perhaps even more clear.

REMEMBER

Screens are addictive. This seems clear, but whatever you call it, there can be a compulsive flavor to much of your screen use; for many people, it may not warrant treatment, but many may experience enough imbalance their lives for them to take their overuse more seriously. Over the last 25 years of my work in this field, I've seen people joke about their screen use as being "addictive"; nowadays, I hear people saying the same thing, but with a different tone. I hear people being concerned about their screen use, and that of their children, loved ones, and employees. I hear their appreciation that this has somehow gotten out of hand. The good news is that until your devices are implanted and always on, you still have some choice in the matter.

The bottom line is this: The Internet, smartphones, and screen devices are amazing, useful, and fun. They give us access to the world in a completely new way; however, what they do not offer is choice. The choice of how we spend our valuable time still rests with us. Use that choice, or lose it through the potent, albeit unconscious, captivating power of an Internet-connected screen.

I have concern about our difficulty in seeing through the provided veil of endless entertainment and in looking honestly at the Internet's ability to render us *choiceless*. This difficulty is further supported by the Internet and technology industry's business model, which is based on our continued mindless use. Choice is never easy, but when it comes to screens, not choosing is worse.

Chapter **15**

Parenting in the Screen Age

P arenting has never been particularly easy. It requires time, energy, and a whole lot of love and effort. It also requires a lot of knowledge and perhaps a bit of wisdom. But parenting has just become even harder. Well, not just recently with the COVID pandemic, but over the last 25 to 30 years with the Internet becoming a daily part of our lives. One of the things I tell parents when I do consultations, treatment, or coaching is that their children, teens, and young adults have much more time and energy than they ever will, so effective parenting around Internet, screen, and technology issues requires working smarter, not harder.

As I discuss in Chapters 2 and 3, children over the last 25 to 30 years were raised with the Internet, and more recently with smartphones. This is essentially a new world, as the Internet didn't exist before. Young people know this new and exciting technology frontwards and backwards. They understand intuitively how it all works, and often have a much better understanding of how to use, overuse, tweak, and hack these devices than their parents. The reversal of historical boundaries, where kids now know more than their parents, creates some unique challenges for parents trying to stay on top of setting limits, rules, boundaries, and consequences around Internet and screen use. This chapter is meant to help parents with those challenges.

The challenge for parents is to first understand how addictive this technology is, and second, to be able to set sustainable and healthy use parameters for their children and adolescents. This is not an easy task in an age where smartphones are like a clothing accessory and a child or teen is glued to their device lest they miss a Snap, Instagram Story, social update, direct message (DM), or text from a friend, or they fail to see (or respond to) a *like,* photo, video or comment on a social media post.

Parents today have some advantages over parents 10 to 15 years ago: They now know how addictive this technology is, and they understand that it is not just harmless entertainment — it requires limits. On the other hand, screen devices today are far more addictive and compelling. Smartphones are only about 11 to 12 years old as of this writing, and they mark a significant change in the way and amount that screens are used, overused, and abused. For one thing, they appear to be the dominant Internet access device for younger users. There are numerous indicators that suggest the effect of the smartphone on your children's and teens' mental and physical health has been profound, and that most of these changes are not for the better.

In her 2017 book *iGen,* Dr. Jean M. Twenge details the negative trend in our youth's mental health, which appears to correlate with the introduction and adoption of the smartphone. Because this device is portable and essentially private, it acts like a secret portal to everything in the world, but these devices also connect your children to the world of social media, which feels so essential to them. It also seems to have addictive, neurological, and psychological consequences in brain development, empathy, self-esteem, depression, social anxiety, and possibly increased suicidality. Although much of this research is correlational, the sharp changes in our youth's mental health seem linked, at least in part, to these devices.

Following Common-Sense Parenting Guidelines

Everyone can agree that good parenting is not easy. Let's face it: Being a good parent is a lot of work, and it requires time, patience, energy, tenacity, and a large dose of knowledge. It can be exhausting because your children all require attention, as well as time to understand, to communicate with, and to address their behavior. This becomes especially challenging when screen use is involved, in part, because your children have become used to their devices and reliant on their self-medicating and distracting qualities, along with the social media apps they are so involved with. Research suggests that most children and teens (and young adults) use social media as a significant part of their digital day, which makes it hard to discount it (nor should we). It seems likely that kids hang out in cyberspace the way previous generations might have hung out at the malt shop, corner store, or shopping mall. However, the Internet, with all the means of accessing it, is addictive, and your children may fall prey to that addictive pattern, often without really being aware of it.

REMEMBER

Part of the issue is today's social and cultural aspects of screen use. You see your children always holding their phone in their hands and assume that that is normal and therefore harmless — but this is deceiving. Excessive use may be typical in children, teens, and young adults, but that does not mean this use is healthy or balanced. Keep in mind that kids lack the full ability to make good executive decisions up until about their mid-20s (see Chapter 3), so even if they wanted to, they may not be able to fully appreciate the impact of their heavy use. Although very challenging, a parent's job is to lend their brain's adult frontal lobes to their kids, until they have full use of their own! This is tricky because you must do it in ways that are age-appropriate and practical within a family's day-to-day life.

In the following sections, I introduce two basic, common-sense guidelines for handling children's screen time: setting limits as a family and being clear, consistent, and calm.

Setting limits together as a family

When kids are interviewed about some of the issues they face in terms of dealing with screen-time limits and how their parents deal with them, there are some interesting findings. One of the biggest complaints kids have about having limits set on their phones or other Internet access is that they hear their parents criticizing them about their screen use, but then immediately see them picking up their phone, laptop, or tablet and using it. Often their parent will say, "Well, this is for work, or something important." This generally does not fly with kids because in their mind, their social media status update is critical for their social life, and their social life is the most critical thing in their life at that moment. Even if it is not, parents must remember that their children have a different view of their world, as they see it from the developmental stage they are in.

It becomes critical to use age-appropriate and reasonable limits and language that are uniquely matched to your child. Logic may not work. Reason may not work. The idea is to find limits that are reasonably developed with your child and perhaps as a family. When a family sets these limits together and the parents follow some basic technology limits (and use some of the guidelines at home themselves), it is more likely that the children will follow the limitations as well.

TIP

Reasonable use may vary, depending on the age of your child; however, studies suggest that two hours of non-academic screen time a day seems to be a Goldilocks amount. Some recent research suggests an hour a day. My research and clinical work over the years has also found that about two hours a day (three on weekends) seems to be enough to allow for appropriate social interaction, but not so much that they suffer negative effects or potential addiction. Obviously, the older the children, the more screen time. I talk more about setting boundaries and limitations later in this chapter.

REMEMBER

Having family meetings where overall tech use for the entire family is discussed is an effective strategy for managing overuse. The idea of setting up healthy tech-use rules and policies for the household engages everyone in a shared goal to be mindful of screen use and emphasizes the collaborative goal of quality, screen-free time. Whatever limits you may want your children to follow, they are far more likely to do so if you are also being conscious of their use and following some appropriate use parameters yourselves as well.

Obviously, parents must work, and work (just like school) requires use of screens. Parents are also hopefully better at prioritizing life — and better at learning to set screen limits for themselves. During the COVID pandemic, the use of screens has been at record levels for both parents and children of all ages, including college-age children, so obviously, latitude is necessary.

Being clear, consistent, and calm

Many of the parents I've consulted with, or whose children I've treated, started their kids' lives with technology by setting unclear or inconsistent limits. In most cases, their children began using technology when they were very young and were provided with many forms of Internet screens and devices to use. Higher-speed Internet was often provided as well. Typically, inconsistent rules and limits were set on how technology was to be used, and they were also inconsistently enforced. Parents also took pleasure in how knowledgeable their kids were about the Internet and technology, and in a sense, this pride often blurred their vision in seeing the damage some of it was causing.

It's hard to set limits, and even harder to be consistent; consistency requires staying on top of your child's use, and nowadays, with everyone having multiple portable and Wi-Fi-based devices, it's nearly impossible to manage all this without some form of automation involved. I strongly suggest that you come up with some form of written plan, with the limits set, so that it can be easy to refer to if needed. I also recommend automatic monitors, filters, limits, or blocks (using software or apps, such as Circle or Qustodio) as much as possible, simply because keeping track of all your child's online activities can become difficult to do, especially if you have children at various ages.

REMEMBER

Being clear and consistent is important, but the most important thing is to be calm. If you have an angry response when your child is overusing or abusing their technology, you will ultimately be less effective in managing your family's technology; in some sense, anger lets your child know that it is *you* who is out of control, not them. Once you get angry, your child knows that you are off-balance, and once this occurs, they know that all they must do is wait it out and you will calm down and let them go about their screen use. Your anger gives your child power.

Keep in mind that no one wakes up and wants to be an Internet addict or even tries to overuse or abuse their technology. It happens slowly, and often without their awareness, and above all, it isn't personal (easy for me to say). Sometimes your strong emotional reaction to your children's addictive behavior may be a response to you feeling responsible for their addiction, and sometimes this reaction can be expressed as anger. Your kids may also lie to you, often if they are doing something they know you will be upset about. Lying can be upsetting to you as a parent, as it can imply that there is a purposeful attempt to deceive you; it probably is not that personal, but rather, more a function of them trying to protect their pleasurable or addicted behavior.

Setting Boundaries and Limitations

Boundaries are part of what makes life work. They are necessary to keep things moving in the right direction, and without boundaries, everything becomes vague and unhealthy. The idea is to be clear about what is expected and what is not. Keeping boundaries vague allows for misinterpretation and confusion, and when you're managing something as pleasurable and addictive as screen use, clarity is critical. The following sections provide guidance on setting screen-use boundaries.

REMEMBER

The funny thing about Internet screen use and children is that parents often have some of the same problems in limiting their own use as do their children. If you want your digital screens (computers, tablets, smartphones, TV, and the like) to be used in a moderate and balanced way, you must model the same behavior! The old "do as I say, not as I do" approach won't work. One of the biggest reasons you may struggle at home with screen limits is that your children may feel you aren't being honest; in other words, they may feel that you're saying one thing and doing another. This can become further complicated if you're working and attending school from home. Trying to achieve balanced screen use for your child can become challenging because you (the parent or loved one) may be overusing your screens as much as your children, and you may not be aware of it.

Establishing low-tech days

What is *low tech?* Low tech simply implies that you identify a day to use less technology, or no technology, than on normal days. It's like eating a limited diet or fasting for a day, only here, you're using only the most necessary technology and leaving the additional myriad of screen use options for another day. Sometimes days like these are tolerated well (and can be good for everyone in the family) because it is only a day with less technology, and surprisingly, people like it better in some ways. It's hard to set a limit defining what is low tech in terms of time, but my suggestion would be *as little as possible,* and certainly that would include no social media, games, videos, and the like.

At the Center for Internet and Technology Addiction, we developed a *TrueTech* values exercise to help people define their personal values regarding how they spend time on their screens. The idea is to begin to develop a balanced approach to using screen technologies by considering what you value in your life. Generally, we've found that most people have never looked at how their values are associated with all the time they spend on their screens, even though they are spending large amounts of time on their devices. The idea is to make both your *values* and your *use* a more conscious choice, as opposed to an unconscious one. We discuss this topic at length in Chapter 14.

A *TrueTech* example exercise would be to identify three core values and determine approximately how much time is needed to manifest these values in your life. Then look back at the time spent on your screens and evaluate whether you've been attending to your screens at the expense of your true values.

Undergoing digital detoxes and tech reboots

A *digital detox* is exactly what it sounds like: the removal of some or all digital screens for a set period. There are no hard-and-fast rules about how much time is involved or whether you include all access to all devices (although the more devices the better, and the longer the better). With a detox, you set a period where you remove the ease of access that is so common with modern technology and screens. The idea is to create an interruption in the autopilot of your (or your children's) tech use, and in this sense, something becomes better than nothing. This helps facilitate a neurological reboot of the reward circuits in the brain.

The only exception to this rule might be if there is an active Internet addiction that has been diagnosed by a doctor or therapist. In those situations, I generally recommend a more extended detox for several weeks, with no access to any screens if possible (but you would want to be consulting with the doctor or therapist organizing the treatment). With school now being virtual, especially during the COVID pandemic, it's important that some access be allowed, but I would recommend that a detox include social media, video gaming, and other forms of screen entertainment. Certainly, access to problem content areas such as video gaming, porn, social media, YouTube, and others would be essential.

REMEMBER

Ideally, if a detox is being considered for your children, it's important that you involve a doctor or therapist who is expert in dealing with Internet and technology addiction, along with a qualified IT expert. Don't attempt to do an extended detox without some professional guidance or support. See Chapter 13 for more information on seeking professional help.

A tech reboot literally involves turning off all your technology for a specified period, and after that set period, starting over, usually from scratch, with a focus on less technology and a reordering of the types and amounts of screen tech that are used. A reboot can be something the family can do as well, although it does take some planning. Again, a reboot is easier if you have professional guidance and support, especially because there can often be strong reactions on the part of the addict or over-user.

Declaring Tech Sabbaths

A *Tech Sabbath* is a planned day or evening where no technology is used. I know this sounds extreme, but I have done this on occasion, and I have to say it is refreshing and amazingly relaxing. People don't realize the amount of stress they create for themselves by always being available on their computers, smartphones, and devices — without a break. Think about it. How long has it been that you have not had your screen device nearby and easily accessible? People are so used to constantly using their screens that they cannot even remember (or imagine) when they did not have access to the Internet. And based on where technology is headed, it seems there will be little opportunity to be off-line in the future (I discuss this in detail in Chapter 16).

There is a commercial on television that talks about finishing the Internet, the joke being that there is no way to ever finish anything on the Internet. The very nature of the Internet is that there is always another place to go, another website, link, hypertext, or back link — forever. Advertisers, businesses, Internet service companies, video game developers, and smartphone providers have discovered the *endless* quality of the Internet and use it to build their businesses; this ensures that your eyes are always on some screen. A Tech Sabbath puts all of that on hold for a brief period. It gives your mind and body a chance to take a break and lower your physiological arousal enough to remember that you are not a tech zombie with a cybersoul. People were never designed to be in a constant state of availability, readiness, and arousal, and the longer-term effects of this "always on" state remain to be fully seen, but we know it contributes to Tech Stress Syndrome (see Chapter 10).

REMEMBER

There is no off button on the Internet, so you must install one for yourself. However, it takes a lot more effort to turn off the Internet than on. The low threshold to act by having available screen access certainly contributes to your overuse. This is because the technology industry's screen world is designed around pushing information and streaming, the idea being that there will always be a constant flow of data, entertainment, and information to you.

Monitoring Screen Use

Monitoring Internet and screen use in the home can be an important tool in helping manage your child's Internet use. I consider this the first level of IT tools to help keep screen use at a reasonable level. Monitoring does just that. It only provides a record of what sites and apps your child is going on and the amount of time they are there, but it is important data to have.

REMEMBER

Why is this useful? Most people experience time distortion when they are on their devices. This time distortion interferes with the feedback they need to be able to make changes in their behavior; however, the feedback alone is not always sufficient to make changes in a child's use patterns, especially if their use has reached addictive levels. But it's a start. Sometimes just providing that information to your child (or yourself) can clarify the limits and rules you have set regarding use, and it gives you, as a parent, valuable information regarding the amount of time and how that time is being used or overused online.

Getting help with whitelists and blacklists

Sometimes the information gathered from monitoring is used to set up limits and rules regarding total time spent online as well as specific limits of online content or websites; this is accomplished by using blocks or limits of specific websites and smartphone or tablet apps. (I talk more about setting limitations and boundaries earlier in this chapter.)

Whitelists and *blacklists* are a means of selectively blocking or allowing specific sites and apps. Which one you use depends on which methods make more sense. Sometimes both may be used:

>> **A whitelist** allows a specified site or app, and no other. You can also prohibit specific sites, apps, or content areas, such as video gaming, social media, and pornography, simply by excluding them from an allowed whitelist. In general, it's hard to only use whitelists, because you will always be adding sites and apps that are needed, and this can become complicated and will require regular management and updating.

>> **A blacklist** specifies a site or app that is not allowed or is blocked. It can also include entire content categories such as video gaming, pornography, and social media. Sometimes you use a combination of whitelists and blacklists; this is because with blacklists, if you don't get it just right, your child may find alternate pathways around the blacklisted content.

REMEMBER

Please consider using an IT professional. They can not only help you with the complexity of whitelists and blacklists, but they can also help buffer some of the emotion that is likely to occur with your children when you set any type of allowed or disallowed apps and websites. In my 25 years of working with Internet and technology addiction, I have found the struggle around setting limits to be significant. In many cases, it is serious enough to create major conflict in the home; having a doctor or therapist help organize this approach, along with the help of a neutral IT person, can help calm some of the emotional energy that is likely to occur when you talk about removing or limiting stimulating or addictive content.

Maintaining consistency

REMEMBER

No set of rules, limitations, or boundaries will work if they are inconsistently enforced or maintained. If there is inconsistency, all you're doing is creating a variable situation where your child will simply *wait it out* until you falter. You may be tired or busy, you may be traveling, your child may simply wear you down, or you may want to reward your child for something they did that was positive. You may forget a limit or rule that you have set. It is easy to forget and hard to be consistent. One thing is for sure: *If you treat screens and technology like a rewarding drug, then your children will see screens that way themselves.*

It is worse to set up a rule and expectation if you know you aren't going to be able to manage it consistently. After all, we are all human, and we have a limited amount of time and energy. It is better to be realistic with your personal situation and limitations and not to set rules for which you cannot provide ongoing support.

I would say that the majority of overuse and abuse of screen technology is not an addiction, as it does not meet the medical criteria for a formal addictive disorder. However, there may be substantial impact for excessive Internet use, regardless of whether it meets the criteria for addiction. Any reduction in screen overuse is probably a good thing and should be considered. The less time you are screen-bound, the more time you or your child has for other aspects of balanced living. Toward that end, any positive movement in the direction of less screen time is always a good thing.

Setting a Positive Example within Your Family

When my children were younger and smartphones were still new, they could not understand why I even had one; it was as if the smartphone was a foreign form of technology that was for children, teens, and young adults. Well, that is certainly not the case today, and I, like most adults (I am a boomer), enjoy all the conveniences and distractions that screens like smartphones, tablets, and laptops have to offer. Adult brains are better equipped to manage the addictive potential of digital drugs, but adults often overuse and abuse these technologies, just as their children do.

REMEMBER

So how do you set appropriate and healthy use limits when you yourself are not doing the same? First, be honest (especially with yourself). To have some credibility with your kids, it's important to say that you realize that you may also be spending too much time on your screens. Do not say that you can handle it. Just let them know that you are aware of how addictive these devices can be, and that you want to help them set better limits; this is also a good time to suggest that perhaps the whole family might need to adjust their screen use as well. The following sections can help.

Starting with small changes

So, here is where you get to make some changes in your own parental screen use, and as I mention earlier, this will build credibility with your children and help you understand how difficult it is to change screen use habits. I am reminded daily, as I write this book and monitor my own tech and screen use, how easy it is to scroll mindlessly through several hours of a day and have little or nothing to show for it. For me, it is my anger at how insidious it is and how easy it is to cross the line online. Ease of access facilitates mindless (and addictive) use. Having your children see that this is a universal struggle might help motivate them, but mind you, they will still likely balk at using screens less often — at least initially.

REMEMBER

Changing habits is hard. Changing a habit that has addictive properties can be even harder. Change takes time. The first step is wanting to change, which may be tricky if the situation isn't dire. There may be less motivation to make difficult changes when there are no serious consequences for not making those changes. The idea is to make change a positive experience within the family and a challenge you can all participate in some way.

TIP

Any small change you can make for yourself or your children can help reduce the autopilot that keeps you on your screens for more time than you want or need. Here are some ideas:

>> Disable the autoplay feature on Netflix, Amazon Prime, YouTube, or any other video platforms.

>> Turn off as many notifications as possible on your phones and tablets.

>> Limit the number of invites you receive from your devices asking you to check in and get caught in a tech rabbit hole.

>> Get rid of any email lists and junk email subscriptions.

>> Simplify the total number of apps you have on all your devices; keep only what you really need and use regularly.

REMEMBER

Everyone can change. You are hardwired to learn and change until you die. Your brain allows you to change patterns and habits if you do something differently for a while. There are no bodies without minds, or minds without bodies; you are what you do, and because of this, you can change who you are by changing what you do.

Trying bigger tips

Research shows that most of the time that people spend on their screens is not for productive (work or school) reasons. People just play around a lot and scroll for hours on end, jumping from photos to social media, to texting, to sports scores, to news, to whatever and then back again. This tech loop can eat up hours of time, and before you know it, it is time for bed.

Kids may lead the way, but adults are doing the same thing, although perhaps to a lesser extent. My advice in the following list assumes that your child is not addicted but may suffer overuse that many experience today. Chapter 14 also outlines some suggestions and ideas for managing an Internet addiction:

>> **Before you look at your screen or device, ask yourself whether you really need to.** Stop and wait for a moment. Do not just reflexively click and look. The idea is to break the automatic process of picking up the phone, or tablet, or another device without thinking about it. Show your kids how you do this. Talk to them about it after you've been doing it for a while. The idea is to change the general assumption that it's "just a screen," "harmless," or "only a video game."

TIP

>> **Give yourself an amount of screen time (non-work or school) that you want to use.** Be generous, but whatever time you choose, you will need to follow for a while. My general recommendation is to shoot for 2 hours a day for nonproductive screen use. That is still up to 14 hours a week, but it is probably hard to shoot for much less at first. Remember that any time you spend on screens is time you are not spending on other (possibly more valuable) things.

>> **Be critical of what you're doing online.** Really ask yourself whether that latest app, website, or game is helpful, or whether it is simply taking up a lot of time. It may be cool or clever, but it may take more of your time than it gives back. An app or website is useful only if it saves you time and adds quality to your life, as opposed to just passing time or being a cool idea.

For example, banking apps are great. They allow me to do what used to take two hours in five to ten minutes. That is a technology win. I wish I could say that for all the apps I have. Many, if not most, do not save me much time, and they pull me in with the idea of something newer and better, but all I end up with is yet another distraction.

Ask your child about the apps on their phone. Ask what they like and why. Discuss what they use their phone apps for, what they like about them, and what they do not like about them. Same for social media: Ask open-ended, nonjudgmental questions. Be open-minded. Have them tell you what they like most and least about their smartphone and other screens.

TIP

>> **Create a time bank.** Buy a few rolls of pennies. For every minute you don't spend on your screen (that you normally would), put a penny in the bank. You won't get rich, but you will see (graphically) how much time you were wasting. None of that wasted time ever comes back to you. You can also use jellybeans (my favorite).

>> **Invest in a bunch of alarm clocks.** Get rid of the smartphones in everyone's bedroom at night. It is a terrible habit, and it is probably responsible for more lost sleep than just about anything else. This is beside the fact that a stress hormone (cortisol) is expressed in response to the phone being near you, which then becomes prohibitive of good and restful sleep.

» **Start using the phone call feature more.** Rely less on texting and more on calling or FaceTiming people. Also try to spend less time on Snapchat, Instagram, Facebook, texting, WhatsApp, Twitter, and TikTok. The idea is about communicating with another person, if that is what you're trying to do. If you're just fooling around wasting time, then that is fine, but be aware of the difference. Texting short bits of information may not be communicating fully. Texting is based on a shorthand and expedient form of communication, which leaves a lot to be desired. It's functional but not complete. It does provide a baseline of conversation, but little else. Decide what your goal is for communicating.

» **Try turning your phone black and white (gray scale).** I mention this idea in Chapter 14. This is a great way to make your phone a lot less interesting. In general, you can expect about a 30 percent drop in smartphone usage simply because it is less rewarding to look at. It seems the brain likes color (that is why you don't have a black-and-white TV anymore).

» **Use the same approach that several industries have used to *increase* your device use: gamification.** Create a game that rewards you for less use of your devices. Track your use and find another reward (a special dinner at a restaurant or a trip you wanted, or something you wanted to buy or that your child might want). The idea is to create a game that rewards you or your child for using screens less. Consider creating a challenge with your kids about who can use their phone the least, and have the winner receive some type of reward.

» **Consider installing software or apps that monitor how much screen or smartphone time you or your child consume.** I mention earlier that I like Qustodio and Circle, but there are many others as well. Screen Time Challenge, Opal-Mindful Screen Time, Kaspersky, and Norton Family are some others. This takes you out of the equation of being the gatekeeper of the statistics; you or your child will also learn to budget how much time you use and potentially help you develop more mindful use of technology. It also decreases the potential for arguments and conflicts between you and your child.

REMEMBER

» **Never use any screen one hour (60 minutes) before bedtime.** I cannot stress this enough. Viewing screens changes circadian rhythms and sleep patterns and increases the risk of inadequate sleep, which many teens (and some adults) already suffer from. There is some evidence that the blue screen light also changes arousal levels and other brain functions. I discuss this in detail in Chapter 10.

REMEMBER

All the ideas and suggestions in this chapter can be mixed and matched; any change, no matter how small, is a positive change. Above all, do something different, as more of the same will likely get you more of the same.

Chapter **16**

The Future of Internet and Screen Addiction

You don't need to be a futurist or technology expert to see that digital and Internet-based technologies will play a huge part in your future going forward. Whether that is good or bad is perhaps too simplistic a question; how screens ultimately affect your life will likely be a mixed bag, with many pros and cons. One thing for certain is that, based on the level of intrusion into your life, if unchanged, your screen use will probably consume a significant portion of your time and energy.

It's also likely that we'll continue to find new neurobiological and psychological evidence that shows the impact of these technologies. We are only now beginning to see the results of what a lifetime of Internet screen use does to people, and how that impact translates into negative health and well-being effects.

Noting That Screen Technology Isn't Essential

I have a unique advantage as I write this book. Obviously, I've been researching, training, and treating issues regarding Internet and technology use, abuse, and addiction over the last 25 years, but I've also had the opportunity to experience being what I call a *digital half-native.*

A digital half-native is someone who spent the first half of their life (or more) without computers, the Internet, and digital technologies. In my case, a good part of my life was spent without what are now considered to be essential forms of technology. This is an important point. The fact that people consider certain technologies such as the Internet, smartphones, and Wi-Fi to be *essential* can be misleading. If they were essential, to my way of thinking, they would be more like running water or indoor plumbing; people somehow convince themselves that they cannot possibly survive without (or with less of) these high-tech tools. But that is just what they are: tools. They make life easier in some ways, but mainly, they make life *different* and in some ways *harder.* You may think life is easier, or perhaps better in some ways, when in fact you have simply readjusted your life to the idea of living with this new high-tech lifestyle. I am no Luddite, and as I've said, I love technology and use it all. I also know Internet and screen technologies are not going anywhere, but if I had to compare the overall quality of life before and after the tech revolution, I would not say it's better.

I completed most of my professional education, including internship and residency, without a personal computer, Wi-Fi, a laptop, a tablet, a smartphone, or the Internet. I'm not bragging, but some people think that would now be impossible. The psychiatry residents and fellows I trained could not imagine going through their education and training without these technologies. When I would give lectures on Internet addiction, I would ask them to put their phones away so they could pay attention. People believe something fundamental has changed in human evolution that has made all this Internet technology somehow essential — that we have evolved to a cyborg-like amalgam of human and computer. The truth is the human race has not evolved neurologically all that much; just our expectations have.

REMEMBER

The first step to being less dependent on your screens, and not being controlled by this powerful (and intrusive) technology, is to remember that people *can* and *did* live without it. You need to remember this daily. You must remember every time you feel as though you cannot live without your screen technology that you really can live with less of it. You can *use* the technology, but not be *used up* by it. I call this maintaining a *healthy tech perspective.*

Being a digital half-native, I can compare life before and after the digital and Internet revolution. Sure, many things are easier, and some things are more convenient. But people are now spending four to seven hours a day (or more) on their screens and devices, in addition to using them at work and school. They have given up so much of who they are for a purported improvement in life that they have forgotten to ask themselves whether they really feel better for it — or whether they feel happier for it. People need to stop and ask themselves whether they feel more connected to the world and those they love.

REMEMBER

I love the promise of technology, and I love how science can improve people's lives. Science is dedicated to acquiring knowledge and moving humanity and the world forward. It embraces the possibility of something better, but sometimes it is not very good at knowing how to temper that growth within the pragmatic realities of balanced and connected living. I would make the argument that Internet and screen technologies, including the smartphone, do not appreciably improve people's lives on qualitative levels, nor do they move the needle of humanity in a decidedly positive direction.

Seeing the Issues That New Technology Will Bring

You must remember that sometimes, more may not be better. More stuff, more tech, and more connections to more people may not equal more quality relationships and more free time in your life. Taking more photos with your smartphone and then sharing them with your 500 online friends doesn't create intimacy or connection. It eats time and energy, and it removes you further from experiencing the moment of where you are now and whom you're physically with. Posting the things you do makes your life a show to be projected and consumed, not a life to be experienced. Stop and ask yourself: How many of those photos and videos do you look at? Does recording and sharing your life on social media help you live it more fully? If the answer is yes, then continue what you are doing; if not, take a chance and try it a bit differently.

Tech companies are just that — companies. They are in business to sell technology products and services (or advertising to other companies). Tech companies want you to buy newer (better) technology, and they sell it to you based on the promise that newer is somehow better, and that if you buy the latest version of this device or service, you will somehow improve the overall quality of your life. The idea that's sold to you is to always be doing at least two things at once, and one of them involves your phone or tablet. Their goal is to create a seamless use of the Internet in your home, car, workplace, school, on your vacations, indoors, outdoors, and while you're walking your dog or with your baby. The idea that there may be a time where you want to be off-line seems almost sacrilegious.

REMEMBER

Better and newer devices, faster speeds, bigger screens, more features — they never equal a happier life.

Facing a loss of freedom

WARNING

How can all these technologies and screens equal a loss of freedom? After all, they are advertised as being all about freedom. When was the last time you felt free enough to leave your smartphone at home, though? All this tech is about lost freedom in two ways:

>> In the amount of time you devote to researching, buying, using, managing, and maintaining all these technologies

>> Through all the data and information you'll be sharing with and providing to corporations that want to sell you things or sell your demographic data to someone else, who can then sell you things

I don't think this is a conspiracy, but rather business responding to potential opportunity; capturing your eyes and attention means big economic opportunity.

Realizing that the latest may not be the greatest

WARNING

There is a new stress disorder that we're seeing today. *Tech Stress Syndrome* (which I mention in Chapter 10) is not just about spending too much time on technology, but also about having to be ever-responsive to the latest forms of screen technology advances. This requires constantly having to re-tool and master new devices, operating systems, software, apps, and the ergonomics of how to operate the latest devices and app options — not a huge deal, but it is another time suck. And that's just it — all the screen devices you use add up in chipping away at your time, and they all do this with the implicit promise that somehow, it will make your life better. It would be fine if this occurred once in a while and you were good to go. But that is not the case. This is a continuous and ongoing process.

Because I've had the opportunity to see the amazing changes in technology over the last 35 years, I can appreciate how much better devices have become. It is true that the iPhone was a game changer and was a vast improvement over previous versions of smartphones or personal digital assistants, but this improvement comes at some expense, and not simply from a financial perspective. Rather, it is in terms of the amount of time and energy consumed by your devices and with all the new things they can do.

People tend to forget that mastery of all this wonderful tech takes time, and that overusing it wastes even more time. The message that newer, better, or faster technology will somehow translate to a more productive or happier life is false. This idea, that better living is always achieved through better technology, is not true; if it was true, people would all be very happy these days, but you only need to watch the news or spend a day seeing what I see with patients to know that people are missing the mark. Technology never really improves humanity; rather, humanity is improved through humane behavior that sometimes may utilize technology while not being dependent on it.

I recently gave up my iPhone 6S (it finally died, and I had to buy a new one). If you know Apple products, you know that was six generations of iPhone ago. I prided myself on using older tech — not for some philosophical reason, but because I really did not want to dedicate more time to mastering a new device, and I also knew that in the long run, it would essentially be like the one I had, and my life would be no better with the latest iPhone. Well, now I have an iPhone 12, and guess what? Nothing has changed (except I have a bigger cellphone bill). This is what choosing your *tech-time expenditure* is about: weighing the benefits versus time commitments and costs. I do love new gadgets and cool features, but I also know that in general, my life is rarely improved by them.

The latest is not necessarily the greatest. It's just the latest. And if you're considering the advantages of the newest or latest (aside from the financial costs), think about all the tech you have now and whether your overall quality of life or happiness has appreciably improved because of this stuff. I know mine has not.

This is not to say that having faster Internet access or the latest smartphone is a bad thing; the Internet is a positive tool, but it is a tool, not a lifestyle. For example, when the Internet began its idyllic rise, people were told that it would revolutionize education. It has not. Test scores have not risen; in fact, they have fallen. During the COVID pandemic, where much of the educational experience has moved to Internet screens, online learning has proven a poor substitute for real-time learning. Let's face it: You still need real-time connection to help stimulate learning. I recall that when I recently went back to school for a fellowship in psychopharmacology, much of it was online, and I found the experience flat as compared with a face-to-face experience.

Identifying problems with 5G and beyond

You cannot watch television today without seeing ads for 5G mobile service, telling you why you should switch to one of the myriad of companies that offer the best, fastest, most mind-blowing cell service. Well, aside from the fact that having faster smartphone and hotspot service may *not* change your life (as they say it will), there are some potential issues with 5G technology and concerns about electromagnetic radiation near human tissues.

I am not an electrophysiologist, but I know enough about radio frequency energy and biology to know that there can be problems with too much exposure. I don't want to make claims that I cannot scientifically back up; I only want to note that there is still much we do not know about all of this technology, and when all is said and done, I am not sure that having lightning-fast cellphone service will be as important as people are being told, and it will likely increase the smartphone's addictive potential.

In Chapters 2 and 4, you find out that the faster an intoxicating substance or behavior is processed by your nervous system, the more addictive it is. This makes sense, in that the association between the stimulating substance or behavior and shorter response latency supports a stronger neural relationship. The brain creates a stronger association when the Internet's digital drug-like effects are quickly followed by changes in dopamine levels, mood, and so on.

Faster access via 5G, or other means of Internet connection, will likely lead to more Internet and technology addiction. From when I began treating this issue 25 years ago to today, I have seen a 20-fold increase in the amount of Internet and technology addiction and overuse cases that I am being consulted on. It is much easier to get access, so it stands to reason that addiction potential would increase. Internet addiction is always more likely with a faster ability to *pick and click*.

Why do people think faster is better? Okay, I get it; when I first got the Internet (way back in the dial-up days), it took three minutes for a photo to scan onto the screen, and now it's on the screen before you can lift your finger off the keyboard! But how else is that better? I know that in the past, people wasted time waiting for stuff to come up, but they're now wasting far more time because higher Internet access speeds increase the time we spend online and our addictive response. Addiction equals a lot of wasted hours. Sure, you're getting a lot of information and content a whole lot faster, but you end up wasting any added time efficiency and benefit by spending so much time in front of your screens. Simply put, faster isn't better unless you manage your total time and attention to your screens.

The tech industry implies that with faster Internet speed and better devices, you'll have more time to live a quality life, but the opposite is actually true. All that speed and accessibility will reduce how much time you attend to the rest of your life. The idea that more tech gives you more leisure time is patently false. People have less leisure and social time today than ever, but they do have a lot more stuff.

Looking at the Tech of the Future

Society is in for some significant changes in the future when it comes to digital technologies. Many of these changes are exciting and offer much promise for people's lives — but they are not without a tradeoff. I am excited about all the possibilities, as I love technology and all the compelling things it can do, but I also know it is not that simple. This is like being excited by that amazing piece of cheesecake that has you anticipating a great experience, but unfortunately this does not negate the unhealthy calories that come along for the ride. The following sections cover what I think is coming in the next 25 years.

WARNING

The Internet still has an off switch, but I fear that in the future it won't. I believe that in the future, people will essentially lose the ability to turn the Internet off because it will be integrated into so many devices and services and tied in to so much of their lives. There will be no off switch because there will be seamless integration into everyone's home, office, car, and personal electronics. Even now, when I come home, I have audio access to the Internet through my Amazon Alexa (yes, I have one of those). I leave work, where I am hyperconnected; in my car, where I am Bluetooth-connected to my phone and the Internet; to my home, which is partially connected to the Internet. Yes, I do have a love/hate relationship with technology because I see so much potential, but I am continually reminded that it's just too easy to get sucked down the rabbit hole, and this requires that choices be made.

Beaming the Internet everywhere

Between cell towers, micro cell towers, microwave towers, 5G mini towers, cell repeaters, hotspots, Wi-Fi, satellite Internet systems like Starlink, and locally networked routers, in 20 years (or less), you won't be able to go anywhere on earth that doesn't have Internet access. You will never be able to get away from it all, because it will all come with you. There will be little opportunity to turn off access and connection because you will feel compelled to use it because it will be available and everything you do will be linked to the Internet. The knowledge that you *could* be available will motivate you to *choose* to be available.

The Internet as you know it today will be almost unrecognizable. It will be everywhere. Everything in your home, office, and car will be connected. This will seem normal and, at times, even helpful. But if you now feel too accessible and too available, in the future, all your digital devices (and everything that can be digital will be digital) will make you virtually unable to hide. All this stuff will require attention, programming, password resets, tech support, and learning and relearning how it operates.

No matter how intuitive all these devices are designed to be, they are *never* always simple. You'll be running around from your refrigerator to your lawnmower, to your lights, to your water system, to your HVAC, to your vacuum cleaner, and then to all your appliances — making sure everything is connected in the way it is supposed to be. But that is just it. All this interconnected and linked stuff is not always that simple to operate, and there are always glitches, problems, and tech support that eat up your precious time. Why do you need all this stuff to be connected? What is the net (pardon the pun) benefit? By *net benefit*, I mean that when you subtract all the time you spend from the promised convenience and life improvement, what do you have left? And God forbid an Internet outage!

Wearing your tech

The smartphone in its current form is clearly a transitional device. It is large and heavy, and vulnerable to getting lost, stolen, or broken. It is a miracle of engineering but is not practical in many ways. Wearable technology, in the form of either glasses or jewelry linked to a larger processor device or uplinked to the cloud or a satellite (perhaps directly to 5G micro towers), will be the next, but far from final, step forward. The iWatch from Apple is a basic version of this wearable technology.

If you think you have no break from your technology and screens now, wait until it's strapped to your face, on your body, or part of your clothing, and you are literally never apart from it. There will be an *off* switch, but the default will be always *on.* You and the device will essentially merge into one unit. It will become nearly impossible to ignore notifications or the constant stream of information that will be pushed through based on everything known about you. Some of this will be useful, but I expect much of it will not be.

Technology already exists that allows LCD screens to be foldable and bendable, and as soon as this is perfected, they will be able to weave screens into the fabric of your clothing, so that your clothing will essentially become the screen through which you see your technology. The cost of this process will undoubtedly go down over time, but what it essentially means is that if you are dressed, your screen will be with you. Add to this the ability to use flexible solar cells that can help power your *smart-garment,* and you'll be a walking television station and hotspot.

The goal of all Internet-based screen technologies is to keep your access available — your eyes always close to a screen.

Implanting devices

Dr. Eric Leuthardt is a neurosurgeon who wants to connect you to the Internet with a brain implant; he believes soon we will be allowed to insert electrodes into the human brain so we can communicate with computers and each other. Elon Musk is actively working on a project called Neuralink, which he hopes to connect people's brains to the Internet; he describes that the procedure will be as simple as LASIK eye surgery. Musk has already invested $100 million into this research. These stories are not science fiction, but are from news stories appearing in 2017 and 2019.

Implanted devices will be the next step after wearables (see the previous section). Even having a screen woven into your clothing, worn on your wrist, or fabricated into your glasses will become relatively inefficient compared to an implantable device. This idea is already being researched — to eventually implant direct access to the Internet into your visual and auditory cortex and sensory processing centers of your brain. Though perhaps future based, this could eventually become a reality. The two news stories at the start of this section are only a sample of some of the work now being done in this area. This is your last opportunity to avoid joking about becoming a human cyborg because the Internet will be connected to you, just like it was in *The Matrix.*

This may sound like the plot of a dystopian novel (in fact, there have been several of those), but numerous individuals and companies are interested in having this technology, or some aspects of it, come to fruition. It may be a few years off, but can you imagine having direct access to the world's population without having to worry about a device or a battery to charge? The human body could supply the small electrical current necessary, and using neural stimulation, you would see and hear as if you were directly looking at or listening to the information. This makes today's virtual reality look like Alexander Graham Bell's telephone.

You could connect to people not just on a contact list, but based on physical proximity and specified life parameters. There would be chosen (or assigned) parameters that would facilitate remote connection to others based on those criteria; this would be the newest and most efficient form of social media networking yet, but with some real privacy issues.

As an interim step, the cellphone transceiver, or parts of it (what you now hold in your hands, less the battery), could be implanted just under the skin with an external rechargeable battery that could be recharged through induction, just like medical devices are charged today or how you charge your cellphone now. This implant would interact with either earpieces or eyewear for the output function, or eventually, through direct neutral stimulation.

Tackling Increased Internet and Technology Addiction in the Future

In the future there will be more Internet and screen addiction because people will be using their devices more often, with faster access, and in more parts of their lives. The faster you get access, the faster the click and the resulting content, the more addictive the whole thing becomes. What can you do about this? Find out in the following sections.

Life isn't found on a screen

When was the last time you saw kids riding their bikes in the neighborhood? That simple and fun activity has essentially disappeared. Between parents being afraid to let their kids outside unsupervised — and the availability of so much stimulating content (including video games) on their devices inside — it becomes a no-brainer. There is no *outside* anymore. No reason to spend time outdoors. No reason to explore, because you can do it all on YouTube, Netflix, or off-planet in the latest video game.

REMEMBER

You are the person who can still choose. You still get to decide (in some ways) what you are going to do with your Internet technology and with your screens. You can choose to spend three hours surfing TikTok, or you can choose to go for a walk or read a book instead. That sounds simplistic, but solutions for too much tech and too much screen time are simple — although they can take some effort to carry out. The choice is easy, but you must find alternatives that provide healthier options for your time and the ability to reclaim lost parts of your life.

Instead of spending 12 hours a week on social media, spend 2 hours a week talking to some friends on the phone; having a cup of coffee; taking a friend to dinner; trying your hand at painting, drawing, singing, or building a puzzle; or picking up that musical instrument sitting under your bed. Do something you have always wanted to do but never had the time to try. You do have the time; you simply must free it from the clutches of your screen.

Recently, I heard an interesting comment: that if you are bored, it is because you yourself are boring. I think there is some truth to that simple statement. People are bored with their collective selves. They are so tired of staring at their screens, like cows ruminating, that they have forgotten the meaning of self-directed excitement. You won't do anything that will add quality to your life if you are medicated with the power of the Internet. These are powerful drugs that will lull you into complacency and boredom intolerance — they artificially raise your dopamine level like a rising wave, only to crash down to less pleasure from real-time living.

REMEMBER

Life is not lived on a screen. It is lived in the moments after you put down your screen. That might sound almost too simple, but it is true. I make much better choices and decisions about my time when I put down my phone or walk away from my devices. Screens and the allure of the Internet distort the world into appearing to be the size of a smartphone, tablet, or monitor. But the world and the people in it are so much bigger than the Internet can portray. People fear missing out if they are not on top of the latest news, social media, or the myriad of other things they find themselves mindlessly checking every day. The idea that you will miss something important and profound online is inaccurate. In fact, the opposite is true: The life you miss is when you are on your screens.

The growing backlash against tech obsession

In the last few years, I have seen a slow, but growing, movement that questions society's obsession with screens and the Internet. People are now examining their social media and smartphone use and beginning to recognize how addictive they are. More research is being published on the potential dangers of screen use in children and the psychological impact that appears to be linked to excessive social media as well as smartphone use. We now understand the dopamine-Internet connection and how the entire Internet operates like a giant slot machine, and that social validation looping keeps people endlessly posting, liking, looking, and posting again. (See Part 1 for details on the basics and biology of addiction.)

I see a developing backlash building and beginning to push back against the ever-growing intrusion of digital technology on people's health and well-being. This is a heartening change; for nearly 30 years, society has steamed ahead as if the Internet could do no wrong, and the more the better.

I see children in my office who are barely in puberty who literally cannot put their phone down for ten seconds, and I see 12-year-old kids who are addicted to pornography. The sad thing about all of this is that many people accept this as the norm today, and have forgotten what humanity looks like with less time on their screens.

REMEMBER

There is hope, however. We are seeing a growing recognition that too much of a good thing is, well, not a good thing, and parents are beginning to make better choices for their children (and themselves). Many people are beginning to choose their technology with some degree of moderation in mind, and they are starting to recognize that the Internet, video games, and all screen technologies are not simply benevolent devices that can be heavily consumed and do only good. They now see that *humanity* in technology use must be part of a healthy tech-life equation.

5

The Part of Tens

Discover ten creative ideas and challenges for managing your technology and screen use. The idea here is to change small things that can slowly increase your mindful and conscious use of the Internet and all the devices that eat up your time and attention. The key is to make the changes fun, manageable, and consciously integrated into your daily routines.

Find (almost) ten suggestions to help you, as a parent or loved one, engage in helping someone who may be an Internet addict or over-user. Many people reading this book will be doing so to help their child or loved one who seems lost in cyberspace. Many of the calls, coaching requests, and consultations we do are with distraught parents, whose children are spending so much time on their phones, video games, social media, or pornography that their lives have become unmanageable.

Apply ten tips to help make screen use that is on autopilot more of a choice. It's impossible to change an addictive or habitual pattern of behavior without paying some attention to your cycles of use or overuse, and slowing down the frequency and manner with which you use your devices. You can use many common-sense techniques to change your relationship with your screens and, in so doing, find more time, meaning, and productivity in your daily life.

Chapter **17**

Ten Things You Can Do to Reduce Your Internet Use

t's likely that if you're reading this book, you are a family member or loved one of an Internet and technology addict; if you're an Internet and screen over-user (a category many people may fit into), you may find the suggestions I make to be relevant and useful as well. To provide something that is potentially meaningful for both groups, in this chapter I suggest ideas that may be appropriate for everyone to reduce their screen use, as well as for parents and loved ones who are dealing with a child, friend, or family member who may be Internet addicted. I hear heartbreaking stories from families who describe how they feel they have lost their child or family member to their smartphones, video games, online porn, or other screen use; the goal of these suggestions is to help you help those people who have lost their way on the Internet. (Flip to Part 4 for even more information on living a balanced life.)

REMEMBER

I noticed that writing this book helped me use my screens a lot less. I am thankful for this because as I said, I love technology, but I do not love the imbalance all this technology creates in my life. I got lost a few times on TikTok, spent too much money on Amazon, wasted some time on Twitter, and became lost in click-bait news stories on my phone. My desire to give less time to my screens was reinforced for me as I was writing this book for you; everyone (including me) needs to be reminded that they are steering their own technology ships, and that they really *do not* need to use their screens nearly as much as they think they do, nor as much as they actually do.

Take a Hard Look at How and Where You Are Spending Your Time

This requires courage as you take what is called, in AA parlance, a fearless moral inventory; this is a list of how you *spend* your daily time (on and off screens), and not how you would like to *think* you spend your time. Research in social psychology suggests that people often distort their self-perceptions regarding their ideal versus real self, and they often regard themselves as healthier than they really are. In other words, everyone distorts and denies parts of their behavior that are not congruent with how they would like to see themselves. This is very common, especially when it comes to addictive substances and behaviors, because you *sort of know* that you are eating, drinking, or on your screens too much, but it can be difficult to first admit this, and even more difficult to begin to make changes in what you do.

TIP

Be sure to be brutally honest with yourself. Look at your screen use numbers on your smartphone, and look at your overall screen time spent on your laptops, tablets, and streaming televisions. If you cannot figure out how much time you're spending, there are good programs and apps to help you do this, like Circle (https://meetcircle.com/) or Qustodio (www.qustodio.com/en/). You can also talk to an IT person to help you set up some monitors. If you're doing this for yourself, make sure someone else holds the passcodes so you aren't tempted to change your usage. I know I've been surprised many times by my numbers if I am not watching them myself. If you're doing this for a loved one, then make sure you are the administrator, so they won't try to pull the digital wool over your eyes.

Start to look for a pattern of what you are doing online and on your screens. Search for the types of content that seem too problematic for you, and the triggers and situations that seem to kick off your trips to digital wonderland. For some people it's boredom, for others it's FOMO (fear of missing out) on social media, and for still others it might be news, sports, or financial information. For many it will be games or porn. Everyone has content or circumstances (or emotions) that trigger them to use their devices unconsciously and excessively. Stress or using your screens in bed at night might drive you into hours of extra screen time. Although there are many reasons why people overuse their devices, once they sensitize those reward circuits in their brain, they are even more susceptible to keep revisiting the method that keeps them stimulated by dopamine. Like an engine, once you start it, it can run on its own for a good long while.

REMEMBER

Once you identify a pattern — for example, when you are waiting for an appointment and get lost on TikTok or Twitter — you can prepare yourself differently by not bringing the device, deleting the app, blocking it if you cannot help yourself, or leaving the device in the car. The idea is to create the opportunity for alternate behavior patterns to reestablish those neural connections and disrupt those patterns of use and dopamine innervation. You are pulling out the weed and planting a nice flower in its place.

Take a Week to Become Aware of How You Use Your Screens

If you're having trouble deciphering just how much you use the Internet and where you go online (see the previous section), try this: Pick a week and just write down (okay, you can put it in your smartphone this once) every time you look at a screen for fun or leisure. I'm not talking about for school or work or banking, but for YouTube, video games, pornography, social media, Reddit, TikTok, Twitter, Wikipedia, Google Earth, looking up memes, fun facts, or surfing through hours of click-bait while trying to read the news on your smartphone. Be honest. You can also use your cellphone's usage tracking feature; the iPhone's is called Screen Time and Android's is Digital Wellbeing and Parental Controls. My guess is that your usage will be way higher than you think, and probably way higher than you want it to be. It takes courage to do this because, just like getting on the scale, people do not want to see what they kind of already know. The problem is that if you don't do this exercise, you will distort what you *actually* do, and you won't have the needed data (or motivation) to make changes in your life.

REMEMBER

The secret to changing an addictive screen use pattern is to change what you do *automatically* into something you do *consciously*.

List Your Goals for Your Life

These can be long-term or short-term goals, but the important thing is to include the specifics of what those goals might entail. The idea is to get clear about what you want for your daily life and in your future, and how the screen time you currently spend is impacting making your goals happen. In other words, describe the goal and when you would like to achieve it, and be clear about what time commitments it will take to make it happen.

For example: If you want to go back to school to finish a degree and you're spending 40 hours or more a week on your screens (not hard to do, by the way), then you may want to rebalance your screen time, considering the desired goal. This is one of the most common scenarios we see in our center — when a teen or young adult wants to finish high school, or more often college, and they are video gaming 12 hours a day. So, what do you think the chances are that their goal will happen successfully if this pattern continues?

REMEMBER

The idea seems simple, but it is surprising how few people do the math or really address the reality of what they are doing with their screen time; it's as if they have endless time and this available time will somehow expand to accommodate their digital pleasures. It does not have to be perfect, but just a rough idea of your goals, when you would like to achieve them, and what it will take to do it — and most importantly, it's about looking at how your screen use may be blocking your path to achieving your goals.

Write Down the Ten Values That You Hold Most Dear

Values are not the same as goals (see the preceding section). They are qualities and aspects of your life that are important to you — they are not necessarily the things you accomplish or achieve. They can be intangible traits or characteristics of yourself or your life, but they may not always translate into action. Many people do not materialize their values because they do not know what those values are. They don't take the time to think about what is really important to them, and of course, values can change throughout their life.

For example, being married or in a relationship might be a value, along with having a meaningful job, being a good friend, or being a good parent (or child). All your values have a *time signature* connected to them, and they all need time (and energy) to realize. Goals can be more structured and concrete, such as climbing a specific mountain, mastering a dance, learning a new language, or finishing school. Or you can also mix the two, which involves an abstract value (say, accomplishing an intellectual challenge) and a concrete goal, such as writing a book — oh, I guess that was my value *and* goal!

REMEMBER

Anyway, you get the picture. The idea is to be very clear about the things that are important to you. Obviously, you may have more than ten, but ten is just a way to start the list. You can always write more if you want, and there are no right or wrong values, or good or bad goals. And the funny thing about values is that sometimes they are not well correlated with what you're doing in your life, so don't be surprised by that. Sometimes values will be something to aspire toward, and you need to stretch your behavior to come in line with what you value — once you have identified what that is.

Pick One Day to Steer Clear of Your Smartphone

Yes, you heard me right. One day. A whole day without your phone. Turn it off. Let your friends and family know they will have to email you, or call your landline (if you still have one), or call you at work. Treat yourself to one day without texting, notifications, social media, or photographing the slice of apple pie you got with lunch — go one day without the myriad of conveniences you use daily that eat up so much of your time and attention. If you cannot do it for a day, try half a day; if you cannot do that, then try for two hours.

The idea is to push your comfort zone for carrying a screen with you so much of the time. You have probably become accustomed to being controlled by your smartphone (and aren't even aware that you have been neurologically conditioned by it); you have become used to being interrupted all the time and being more attentive to a screen and the people you are Instagramming, Snapchatting, or Facebooking with than the people physically right next to you. The things that comfort us can end up jailing us.

TIP

If you cannot go two hours without your device, then consider breaking your habit by not looking at your smartphone or tablet during the times you would typically use it. Change *one way or time* you use your smartphone or other device. The idea is to change some very small aspect of the behavioral and neurobiological pattern. By disrupting the pattern, you create room to take on new behaviors and receive the psychological reinforcement from knowing you can change — even if in a small way. Small changes now leads to bigger changes later.

Take the Waiting Room Challenge

I love this one, and I use it frequently myself. The smartphone is barely 12 years old, so anyone 25 or older should recall sitting in a waiting room before smartphones existed — you sat and waited. You did not die of boredom because you had no screen distraction. Maybe you read a magazine — remember those? Maybe you did something very unusual by today's standards and struck up a conversation.

Today you would be hard-pressed to find someone not looking down at their phone.

The idea is simple: to take the *waiting room challenge* (it applies to any place you would automatically pick up your phone or tablet). Simply change what you have been typically doing when you are in a situation where you are bored. Welcome the brief opportunity to be bored. Do not pick up your device when you are at a red light, in line at the post office, in a waiting room, or in any situation where you must wait for a few minutes. Stretch your boredom tolerance to a few minutes without entertaining the distraction of the screen; learn to take the moments of boredom as an opportunity to think, relax, talk to the person next to you (if their face is not in a screen), have creative thoughts, plan to do something, daydream, take a deep breath, or close your eyes.

WARNING

By the way, if you think you're being productive by using those seconds or minutes on your screen for something, don't be fooled. Productivity needs intermittent off-times, and the way people live today, they are *never* off — ever. This is abnormal and biologically counterproductive to being ready to work at peak efficiency. People think they are being clever by stealing wasted time and by grabbing those moments on their screens, when in fact they are missing natural and necessary opportunities to rest, refresh, and reboot.

Stop Taking Your Phone into the Bathroom, the Kitchen, or on a Walk

I know that you may have an emergency while you are walking your dog or walking with your baby, but try to live without it for that brief period. People take their phones, tablets, and other devices with them so they can kill two birds with one stone. So, they try to multitask when multitasking does not really exist (see Chapter 4). The problem with doing two things at once is that you are not fully present for either of them! Or you are only half present for each. Do the *one* thing you want or need to do. Walk your dog, walk with your baby, go to the bathroom — take the moment to do what you need to do, and do not divide your attention by doing two things at once. If you're eating a meal or sharing a meal with a friend or family, make sure the phone is nowhere in sight. There is absolutely no reason to have your phone with you during a meal. We survived eating meals without phones for centuries.

REMEMBER

Focus on the experience you are doing, not somewhere or something else. You'll find you get much more from your life when you choose to fully experience one thing at a time, and to be present in the moment.

Delete All the Junk Email You Have and Unsubscribe to All Unneeded Emails

TIP

Clean out your *digital closet.* Delete everything you do not need. Unsubscribe to useless and annoying emails and marketing texts. Get rid of the electronic junk that clogs your mind daily and slows down your ability to focus on the digital information you *must* attend to. There is so much information and input to sift through: emails, texts, social media, instant messages (IMs), FaceTime, direct messages (DMs), phone calls, Zoom, and snail mail. The list is almost endless, and the stream of information and data you need to process is far too much for you to attend to in a healthy and productive manner. Say no to the stuff you do not need. Get rid of all those notifications and junk emails that tell you about all the things you do not need or want; the seconds you spend sifting through each of them add up. Reclaim your eyes for more important matters.

Turn Off Every Notification You Can

TIP

You do not need 95 percent of the notifications you get. They are distractions and invitations to look at your screen and get you sucked into wasting valuable time onscreen. The word *notification* implies that it is something important that you are supposed to attend to, when mostly it is the latest flash or pop-up sale or some other plan to separate you from your precious time and money. Notifications are the *drug pushers* of the Internet, each one trying to coax you to click on something. Just say no, and turn them off!

Put Yourself on a Digital Diet

Think of your screen time as you would any other part of your life. You want to be able to choose how you spend your time and energy based on your values and goals (covered earlier in this chapter), as opposed to unconscious or addictive habits. Setting limits and creating an overall digital use plan can help you manage your non-work or academic use time (see Chapter 14 for full details). The idea is to see your screen time in a *dosed format* where you treat yourself with some *junk time* online based on balanced living. In a sense, this is how television has been used by many people for a long time (although that is changing with Internet-based streaming). It is not that TV was never overused at times; sure, who hasn't stayed up too late on a work or school night watching that favorite rerun, only to regret it when the next morning arrives all too early? But something is different with the Internet, video games, streaming TV, and smartphones — they are more addictive.

In my decades of practice in the field of addiction medicine, I cannot recall treating anyone for a television addiction. I am sure there were some people who might have qualified for this diagnosis, but I never saw it. With the Internet, with the amount and types of available content (very stimulating as well as mundane), ease of access, availability, and portability, we have seen the creation of a new disorder. Internet addiction was born almost simultaneously with the popularization of the Internet, only to gain more momentum as the technology advanced. High-speed access, Wi-Fi, portable laptops, tablets, and especially smartphones — each advance presented a new iteration of a growing problem.

Internet and technology addiction is not actually new; as I discuss in Chapter 2, behavioral addiction is hardwired into the limbic reward circuits in your brain. But the idea that something that seems so harmless and fun (and so productive) could derail and unbalance your life surprises even me. I am constantly reminded of how much time I can lose if I don't watch myself when using screens, and writing this book has given me a renewed awareness of just how insidious Internet screen technology can be.

REMEMBER

Mindful technology and screen use is necessary; the idea that you can use a device and keep accurate track of your time usage cannot be counted on. I've caught myself on more than one occasion lost in cyberspace, and I have to come back to my senses and turn it off. Sustainable use of your screens can be achieved only with a vigilant use of this useful, but addictive, technology.

Chapter **18**

(Nearly) Ten Ways to Help Your Loved One with an Internet Addiction

Many people reading this book will be doing so to help someone they know and love. Much of my day is spent consulting with other doctors and therapists, and helping parents deal with their child's screen use, whether it involves video games (which seem to be the most common screen issue), social media, pornography, YouTube, Reddit, or their smartphone. To get past Internet use problems, the same question comes up: How can someone help their child or loved one get past their addiction to the Internet? This is a challenge for any parent or loved one because they can see the obvious changes in their loved one's life, but often the addict doesn't see anything wrong with their use.

Part of the problem with video gaming, for instance, is that the word *game* implies that it is simply a game, and therefore harmless — that it is nothing serious to worry about, and as a fun pastime it should not create any problems. But nothing could be further from the truth. I have seen video games essentially take over a person's life, where their use is so frequent and pervasive that they are barely functioning in any of the major areas of their life; they frequently experience significant changes in their academics and social behavior, as well as isolation when at home and with friends. Other psychological problems may also be caused or exacerbated by their screen use, such as depression, social anxiety, ADHD, or sexual behaviors.

The video game industry doesn't help either, with an official position that video games are simply entertainment and are never addictive (despite the World Health Organization's recent ICD-11 diagnosis of Gaming Disorder); this is made even worse by the fact that video gaming is now regarded as a competitive sport, and some small number make their living from playing.

How can a partner or loved one intervene? What power does a parent or family member have? How can you best help? The ideas in this chapter can guide you as you help a loved one. (Flip to Chapter 15 for specific tips on parenting in the screen age.)

Communicate a Clear Message of Love and Concern

REMEMBER

Above all, let your child know that they are loved no matter what, and that you can disagree with them and still love them. Be clear that you are not comfortable with their behavior, but do not inadvertently let your child believe that your love and concern for them is permission for them to keep doing a harmful behavior. Do not expect your child to agree or support your opinion regarding their behavior, and keep in mind that you do not need their approval to parent as you see fit. I often see parents asking their children for permission to set limits, rules, and blocks on their devices, and then the child says no — or they throw a temper tantrum. It's okay to do what you feel is best, and to let them know that you are not okay with things as they are and that you want to help them make some changes.

Consider the Level of Help That Your Loved One Might Need

Sometimes all the addict or over-user needs is more information, guidance, and support. Other times, more structured treatment may be needed. If you have tried general counseling and psychotherapy to address this issue, it might be time to arrange for a consultation with an Internet and technology addiction expert. Often just getting some guidance and information for you as a parent or loved one can help the situation. Other times, I might suggest some outpatient (either in person or by telemedicine) sessions with a doctor or therapist who has experience treating Internet and technology addiction to see whether that might be helpful. At other times, more intensive treatment may be needed, such as an intensive outpatient program or residential treatment program.

REMEMBER

When it comes to addiction, there is a continuum of care, starting with information and support at the simplest level, and progressing all the way to inpatient/residential treatment at the other end of the spectrum. Support groups for the addict or the parent or loved one can also be quite helpful (see Chapter 12). Sometimes just getting a consultation or coaching from an expert can help you help your child or loved one get started, but do not get discouraged if your child or loved one does not initially seem motivated to receive your help.

Try to Get Your Loved One to Join in Your Effort to Help, But Take Your Time

REMEMBER

Keep in mind that no one chooses to be an addict, but they do have to choose *not* to be, and this process takes time. The addict or over-user needs to come to a point where they recognize the impact of their behavior on their life, not just that it is irritating or annoying to a parent or loved one. The addict needs to develop some need and motivation to change themselves, but this process can also be facilitated by healthy parental limits and boundaries. This gets complicated when they are living in your home and they are a late teen or young adult, and still more complicated if they are away at boarding school or college.

Sometimes the way you get them interested in changing is by providing information, resources, and research, and after doing so, asking them whether they would like to know more about the issue. You should give them some power to choose their readiness for help, but this should be structured as a *forced choice*, meaning that any choice of the presented options is moving them in the right direction. Psychoeducation and medical information about how the brain works and how the addictive process occurs can be invaluable in increasing motivation and comfort to change — but do not be insulted if they initially reject the information being offered.

Have a Conversation about Your Loved One's Goals, Desires, and Interests

Engage with them about their gaming or Internet use. Find out more about what they like and what they are doing when they are online. Ask them about where they are going on social media, what games they like or do not like, what they are specifically doing in a game, and what roles or activities within the game are exciting for them. Try not to be judgmental — the reality is that most of us have little to no understanding about what our kids are doing online, and we need to.

Talk to them about what their plans are, and when they say they want to be a professional gamer or a game developer, do not laugh or criticize them. Ask them to tell you more about what is involved. Although it is unlikely those things will come to pass, it is important to understand your child or loved one's goals and what is motivating them.

REMEMBER

If you can get them to talk about life goals, all the better; the idea is to help them broaden their view from the narrow world of their screens to something wider and more encompassing. Do not expect them, however, to see the world from the same place you do; there are generational differences and digital native factors, and of course, if they are addicted, they will undoubtedly be skewed in their attitudes. Their viewpoint may be unrealistically positive about technology, but *listen*, even if you do not agree; the idea is to learn what is important to them, even if you ultimately decide to block or limit specific content or the total time they spend on their screens.

Set Limits and Boundaries

Most parents or loved ones have tried setting limits on the screen use of addicts and over-users. Many times, they set verbal limits, followed by the addict's promises of compliance, only to have those promises quickly fade. This is the typical pattern we see: Either the well-meaning parents or family members verbally correct the addict's behavior, or there are half-hearted attempts to use apps or software, or even to remove the router or other equipment to stop their Internet or video game use. Many times, these efforts are inconsistent and quickly forgotten, only to be rediscovered when something goes wrong in the form of poor academic performance or other psychological or behavioral issues, which reminds parents or loved ones that there is still a problem.

TIP

If you want to set limits or boundaries around an addict or over-user's screen behavior, I strongly suggest that you automate those limits; trying to do this manually, as needed, assumes the addict will remember (but you know they won't want to remember), and it also assumes *you* will remember, find the time, have the energy, and be willing to go to battle yet again over them spending too much time on their screens. By automating limits and boundaries, you remove the guesswork, the need to remember, and the emotional tug of war that often occurs around setting limits for the amount and types of use. Many software programs and apps will help you do this, and you can limit or block specific content areas (such as social media, video gaming, or porn), limit the total amount that they access such content, or create broad limits on their Internet device and screen use (including access to the Internet through Wi-Fi, 4G, or 5G). Obviously, it is always better to simply set limits and boundaries and have them politely followed, but my experience suggests that often, it may require an automated technology approach to limit potential conflict and simplify the process.

Use Monitoring, Blocks, and Filters

Blocks and filters are specific tools that are designed to eliminate access to problematic and dangerous areas of Internet use. They are generally helpful if you want to block video gaming, online shopping sites, gambling, or pornography (covered in Chapters 7, 8, and 9). Sometimes it is a good idea to *monitor* their usage (apps, websites, and types of content) to see what the problematic content areas are and their usage times. It is important to assure the addict or over-user that specific messages or personal information will *not be seen or monitored*.

Sometimes it may be necessary to physically remove a device, and you as a parent should check in on your child randomly to make sure they are not hacking through the blocks or on another old device they found in the basement. Internet addicts are typically very resourceful and technically skilled when it comes to computers, the Internet, and screen devices, so care must be taken when making any hardware or software changes; this is again another reason to work with an IT professional to even the playing field.

REMEMBER

Many times, the addict has no idea how much time they are devoting to their screen use, so monitoring can be a good first step for getting data to see the specifics regarding the nature of the problem. This information can be used as a basis for any blocks and limits, should you decide to use them, and it can also provide the addict or over-user with hard evidence that shows exactly what they are doing on their screens.

Consider Professional Help

REMEMBER

Sometimes professional guidance can help. The first step might be for a parent or loved one to get some coaching, or to consult with an Internet addiction expert to offer suggestions. The next step might be to have that expert meet with the addict to evaluate the addict and the level of problematic or addicted use. Sometimes this might be followed with another consultation with the parents to offers suggestions, and lastly, direct treatment might be necessary in the form of outpatient treatment (via telemedicine or in person), intensive outpatient options, partial hospital/day treatment, or a residential treatment program. See Chapter 13 for more information.

Don't Take Your Loved One's Use or Lies Personally

Many of the issues that parents or loved ones raise involve the lies they are told by the addict to rationalize, justify, or distort the addict or over-user's behavior. Sometimes the lies are not intentional, in that the addict is likely unaware of how much time they are spending online or on their screens. I have had video gamers swear they spent two hours a day on gaming when the actual number was closer to six or eight hours. Other times, they use intentional lies to cover up their use and to keep doing what they want (and need) to do.

REMEMBER

Above all, do not take the lies personally. Addicts often lie — lying is necessary to be able to continue a behavior when they feel terrified or afraid to go without the behavior, even when there are negative consequences in maintaining that behavior. The lie allows the addict to keep using their screens, and telling a lie becomes a small price to pay to get that dopamine hit, as it feels like a "survival instinct" to them. Addicts often do things that are essentially inconsistent with their values because the addiction overrides judgment and executive thinking. It is important to have some compassion for the addict, because when they are inside the addiction, they may become captive to the primitive neurobiology of their addicted brain (see Chapter 2).

Above All, Have Hope

REMEMBER

Please have hope. People have an amazing capacity to change and grow; the brain is capable of neuroplastic change and adaptation throughout a person's lifespan, and especially so during their youth. People continue to learn based on experience, and they can heal from mental health and addictive disorders — at times with some help. Addictions are inherently stubborn human problems and can require numerous attempts before achieving a level of sobriety or moderated use; obviously, when it comes to the Internet and screens, a sustainable, conscious, and moderated approach is required because people need to use these technologies for daily living, work, and academics. They do not, however, need to spend ten hours a day video gaming, on social media, or on their smartphones.

People can and do change, and they can get better from addictions; sometimes, an addiction to screens gets better over time on its own, but often it requires some effort to manage the addiction and achieve a more balanced life. One of the most important messages that I provide to parents and loved ones who seek my help is that there is hope for change, and that most of the patients I have consulted on, or treated, have achieved improvement with their addiction or overuse. Addiction is not *cured* because there are relapses or slips, but despite this, most addicts can return to their academics, work, social life, and balanced living, and go on to achieve success. Over time, with appropriate education, parental support, consultation, and, at times, treatment, improvement happens. The most gratifying thing I see is that parents and family members feel they get their loved one back after a protracted absence into the abyss of cyberspace.

REMEMBER

In the field of addiction medicine, we think in terms of *harm reduction,* and that any improvement is positive. If we can get a patient to reduce their screen use by 50 percent, or to remove problematic content from their daily use, that is a win. Any improvement is a success (though a parent or loved one would prefer that the addict or over-user have no issues at all with screens), so some change in a positive direction is always better than none. Many times, small increments of change will be cumulative over time to eventually produce a bigger change down the road.

There are times, of course, when abstinence from specific content areas, apps, or websites might be necessary. Typically, we see the need for avoiding certain apps and websites with the most problematic (addictive) content — such as video gaming, pornography, social media, gambling, and shopping sites — which could still allow necessary Internet and screen access for academics, work, and moderate entertainment.

Chapter **19**

Ten Tips for Overcoming Internet Addiction and Screen Overuse

No simple list of tips (like the ones in this chapter) will change your life by itself, but doing a few things differently can yield the beginning of new habits and changes in addictive behaviors. Behavior patterns are often linked together with triggers, leading to overuse and resulting in physiological and psychological reactions. Changing something in this overall pattern can positively disrupt the entire system — new thoughts can change behavior and changing behavior can change habitual patterns. One small change can slowly lead to other changes in how you use your screens, including the content you consume and the amount of time you spend on your devices. The idea is to change something, with smaller changes later becoming larger changes. (Part 4 has even more guidance on balancing your tech life.)

REMEMBER

No medical or psychiatric diagnosis can be made without a professional evaluation. The tools and tips provided in this chapter are intended for educational and informational purposes only. If you are concerned about your smartphone, Internet, and technology use (or your loved one's), you may want to also consult with a licensed mental health/addiction professional with expertise in Internet and technology addiction.

Set Aside Time and Places for Not Using Screens

Pick times or places that you can experiment with using less tech. For instance, don't take your phone with you on an errand or walk; leave your phone in a place that allows you to do something (even for a few minutes) without your screen. Most people carry their phone *everywhere* they go, and therefore they are more likely to overuse it. Try having a meal without your phone as one of the place settings on the table, and avoid using your device in your bedroom or the kitchen. Kitchens are for eating and preparing food, and bedrooms are for sleep (and, well, other things), but the message to your nervous system is that both rooms can be places of self-nurturance and rest.

Having screens in the bedroom at night can be disruptive to sleep due to the impact on your circadian rhythms by the LCD screen's blue light. It also establishes a neuropsychological and neurobiological habit that alerts your body to be in a state of readiness — this leaves you less relaxed, more stressed (with elevated cortisol), and less able to be present in the moment. (Find out more about the dangers of having screens in the bedroom and the kitchen later in this chapter.)

WARNING

People were never designed to be in a near-constant state of arousal, and having their phones with them everywhere they go contributes to this unnatural way of living. Knowing that your phone can, and will, be delivering notifications, new information, texts, updates, and calls keeps you in a hypervigilant state of arousal. Just seeing your phone (even if it is silenced or off) can elevate your stress hormones!

Years ago, most people had one wired phone in their home, and when the phone rang, they had to run to the room where it was to answer it or make a call. Then everyone got extension phones in other rooms of their home, so they could conveniently make and receive calls. Now everything has come full circle, where people have one phone again and must carry it with them everywhere, lest they miss a call, Tweet, update, Snapchat, Instagram, instant message (IM), direct message (DM), or text. The problem is that they never get a break from being available and always on, and they are *chained* to their devices.

REMEMBER

Walk your dog or with your baby without your phone. Just take in those few minutes to experience life without the added distraction of work, conversation, social media, surfing, scrolling, or anything that takes you away from the moment. People keep looking for digital distractions, and so they never get the benefit of the experiences they are having in the moment. You cannot be in two places at once — you must choose where your motivation and attention are to experience life more fully.

Never Have Your Phone Out during Meals

REMEMBER

The idea is to create healthy boundaries around your Internet and technology use. The smartphone is *not* an eating utensil, and there is no such thing as multitasking. Your child will tell you they can multitask (and you may also believe you can multitask), but the brain can process only one stream of information at a time. You must attend to a single piece of information to fully process it, as your attentional capacity is limited — the more things you do at once, the longer it takes to accomplish each task at hand, and the less comprehension you have for each task.

Stop Using Your Phone as an Alarm Clock

Do you really need an $800 alarm clock? The idea that you need your phone to wake you up is frankly ridiculous; this encourages you to use your phone in your bedroom and to play with it before bed. If the last thing you look at before sleep is your phone, then it will most certainly be the first thing you look at in the morning — and this starts the *screen use cycle* all over again.

TIP

Viewing screens changes circadian rhythms, impacts sleep patterns, and increases the risk of inadequate sleep (which many teens already suffer from). There is some evidence that the blue-screen light also changes arousal levels and other brain functions. You can buy an alarm clock for $15, and there are many other ways you can wake up. I think they still even make clock radios — remember those?

Avoid Smartphones or Screens at Least an Hour before Bed

REMEMBER

Bedtime should be used for sleep or preparing for sleep. When you engage in screen activities, your brain is activated in ways that are the opposite of sleep; neurotransmitters and stress hormones are released from screen use, and it's like stepping on the gas pedal when you should be coasting to a stop for the day. The idea is to create the necessary rest that you crave and need to function psychologically and physically; research shows that most Americans are chronically sleep deprived, and there is evidence to suggest that some of the reasons for this have to do with screen use. If possible, do not use screens an hour before bed, and avoid sleeping with your phone under your pillow or next to your bed.

Turn Off as Many Notifications as Possible and Consolidate Apps

All addictions have triggers, and the Internet and your smartphone trigger you (via notifications) in ways similar to how drug, alcohol, and gambling addiction gets triggered. Notifications let you know something good *may be* waiting, but it doesn't even have to be that direct; it can simply be seeing the phone, computer, or tablet, which then serves as a trigger. The time of day, fatigue, stress, any strong emotion, or simply having a few free minutes can also serve as triggers.

What are notifications anyway? Well, they are invitations to feel a hit of dopamine, followed by a slightly smaller hit of dopamine. They are triggers designed to get your attention, but they do more than that; notifications are pings to your nervous system to signal your nucleus accumbens, deep in your brain, to fire up those dopamine receptors and get ready for that hit of dopamine, which sets up a loop of receptivity for dopamine innervation (see Chapter 2 for more about the biology of addiction). The more notifications you have set up, the more you will be distracted and dependent on your phone. These are like the lights and sounds on a slot machine that let you know that *if you play, you might win something*, but if you don't win, there will be *another* chance.

REMEMBER

Smartphones are the world's smallest slot machines (the broader Internet to which they are all connected is the largest one). They both operate on a variable ratio reinforcement schedule, and as such, you never know what you are going to get, when you are going get it, and how good it is going to be; this is the essence of variable conditioning, and it is very resistant to extinction (which is another way of saying it is addictive). Notifications are your machine's way of letting you know there *might be something you would like to see.* This gives you an anticipatory dopamine release, and then, if it is something you *really like,* it is followed by another lesser release/reception of dopamine.

TIP

On a related note, you should consolidate all your apps on all your smartphones and tablets. Eliminate the ones you don't use. The idea is to limit the number of icons you need to sift through that eat up your time and attention. Make your devices as simple as possible so you only spend the time that is necessary. Many of us become tech junk collectors, making life more complicated.

Install Software or Apps That Monitor Screen Use Time

Consider installing software or apps that monitor how much screen or smartphone time you or your child consumes. This takes the guesswork out of the equation, and gives you the feedback you need to accurately judge how much time either of you spends on screens.

REMEMBER

Because most people distort the amount of time they spend on their screens, they need to realistically learn to budget how much time they spend, and to develop more mindful use of their technology. This can be especially helpful in decreasing the potential for arguments and conflicts between parents and children over how much time is spent on screens. Having some hard numbers can do away with denial and distortion that everyone is susceptible to when it comes to their screen use.

Monitoring can provide you with a starting point for what you are doing and how long you or your loved one is spending on screens, and then provides you with the needed data to help put boundaries and blocks on devices you want to limit. Sometimes, setting a use allotment for total daily screen time can be helpful, while at other times, it might involve blocking specific problematic content such as video gaming, social media, pornography, gambling, or anything that sends you down the rabbit hole of the Internet. Other times, whitelists (allowed apps and websites) and blacklists (disallowed apps and websites) can be set up to simplify the management of screen time.

Set Your Screen to Black and White

TIP

The ability to change the color of your smartphone screen to black and white makes it about 30 to 50 percent less appealing, and so you won't want to look at it as much. The human eye is attracted to color, and all the stuff you like to look at on your phones and tablets looks a whole lot better in color. So, if you want to use your phone less, make it less fun. I can promise that you won't miss it too much. For the most part, with the exception of a serious piece of news or videos and photos of your family (or something for work or school), most things can wait a bit, or you can just see them in black and white (you can always switch back to color at any time).

The constant, mindless surfing and scrolling you do can also be reduced by simply making the screen less stimulating to look at. You won't even be conscious of it, but you will just naturally use it less, because it will be less interesting visually. This is accomplished through the settings on your device, and you can look up how to do this on your browser for your specific device. It takes about one minute to change, and you can always reverse it if you cannot tolerate a *more* boring screen.

Create a "Real-Time 100" List

I introduce this list in Chapter 12. Coming up with 100 things you can do without a screen shakes the mental cobwebs off your psychological flexibility and creativity. The fact is that most people over 20–25 may still remember the time before their smartphones existed, when they had to look for other ways to entertain themselves. Although the Internet has been around for the last 35 years or so, the faster and more portable it has become, the more addictive it has become. Trying to develop substitute activities you can enjoy when you aren't picking up a device can really help — especially because you may have forgotten how to live without your phone or other screen as your first option.

REMEMBER

Nature abhors a vacuum, and having some non-screen activities to fill your time and desire for pleasure will go a long way to reminding you that there is more to life than can be found online. It also allows you to create new neurological patterns so that you don't automatically pick up a screen when you have several other readily accessible options.

Learn to Tolerate Boredom

Boredom is not something to be frightened of and can be your friend. Many people can no longer tolerate boredom, and they have lost the ability to use other inner skills and resources, which have atrophied from the digital fast-food diet they have become accustomed to. Try not to jump onto your screen the second you have a free moment. This immediate act of reaching for your device blocks your ability to access inner resources that spring from the depths of your boredom. If you medicate yourself the instant you have a free second, you never exercise the muscle of being in the present, without the distraction of your screens. If you don't instantly *screen up* when you're bored, you will slowly develop the ability to make other choices in those moments.

REMEMBER

Because you have become so boredom intolerant, even a few minutes of *doing nothing* can feel too long. Ten years ago, it felt awkward and uncomfortable to pull out a smartphone in public — now it's just the opposite. Today, people feel uncomfortable not taking out their phone, as everyone else has their phone out, and it feels strange to not be following today's social norms. Dare to be different.

Never Pick Up Your Phone While Driving

WARNING

Never have the phone too easily accessible while driving, and resist the temptation to take that quick second to do something on your phone while driving. It's better to use the hands-free system in your car to talk, but even this can be somewhat distracting. Teens and adolescents (as well as adults) are very susceptible to distraction and are six to seven times more likely to have an accident when using a smartphone while driving. The teen brain is less well developed in the frontal executive region, where reasoning and judgment reside, and so teens are more prone to poor decision-making than adults — although both are susceptible to distracted driving. In the distracted driving research that I did with AT&T in 2014 and 2015, we found that most of the reasons we use (and overuse) our phones at home or work also apply while we are driving — this is basically Russian roulette with a screen and car as weapons.

Index

B

C

CBT (cognitive behavioral therapy), 225–226

Center for Humane Technology, 219

Center for Internet and Technology Addiction, 218, 219, 270

Center for Parenting Education, 221

Child Technology Test, 194–197

children. *See also* parenting; teenagers

  addiction in

    biological factors, 44–47

    health concerns, 160–161

    lack of experience, 49–51

    other factors, 52–55

  advertising and, 255

  being technologically adept, 305

  boredom, 54–55, 175

  brain development, 36, 45–47

  Child Technology Test, 194–197

  digital devices, 52–53

  lower-tech living, 202

  maybe factor and, 47–48

  multitasking, 66

  relationships, 53–54

  sex hormones, 47

  sexting, 133–134

  social media, 78

  technology, 53–54

  withdrawal symptoms in, 38

Children and Media Research Advancement Act, 255

Children and Screens, 219

Children's Screen Time Action Network, 221

circadian rhythm disruption, 166–168

Circle app, 206, 219

coaching, 242

cognitive behavioral therapy (CBT), 225–226

cognitive defense mechanisms, 253

cognitive symptoms, 148–152

Common Sense Media, 219

communication. *See also* relationships

  cyberbullying and, 82–83

  goals, desires, and interests, 304

  within Internet, 53

  love and concern, 302

  parenting, 268–269

  setting limits, 230

  social media, 78–80

community, video game, 103–106

computer viruses, 114–117

consumer bulimia, 118

contemplation as stage of readiness, 236

convenience, 113

Cooper, Alvin, 60

covert sexuality, 139–140

COVID pandemic, 259

cravings, 262–263

credibility, 108, 274

cyberbullying

  overview, 25–26

  power of, 82–83

  social empathy, 80

cybercoma, 207

cybersecurity

  dating sites and, 135

  defined, 114–117

  overview, 25–26

Cybersex Abuse Test, 192–194

cyberstalking, 25–26, 83

D

dark web, 18, 121

Darnai, Gergely, 165

dating sites, 134–135

deep vein thrombosis (DVT), 160–161

default mode network (DMN), 165

dendrites, 31–32

dopamine
addiction, 28–29, 35–37
anticipation, 35, 38, 63–64
brain development, 36
children, 44–45
defined, 28
depression, 228
lying and, 306
marketing affecting, 117
maybe factor, 11–12
online gambling, 120–121
online investing, 17
online shopping, 118–120
overview, 27–28
passage of time and, 59
positive reward transfer, 158
receptors, 30
regulating, 204–206
sex releasing, 127–128
social media, 75–76
triggers and, 38
variable ratio reinforcement, 68
video games, 101
dosing schedule, 205–207
down-regulation, 34–35, 156
driving and texting, 171–174, 315
DVT (deep vein thrombosis), 160–161
dynamic interaction, 69

E

effortless starting. *See* autoplaying
elevated cortisol, 165–166
EMDR (eye movement desensitization reprocessing), 242–243
emotions. *See also* behavior
anger, 153–154, 262, 269
anxiety, 156–157, 227, 244, 255
depression
dopamine and, 228
medication for, 227
reward deficiency syndrome, 99

as symptom, 155–158
treatment, 244–246
digital detoxification affecting, 204
EMDR and, 242–243
information affecting, 92
managing with video games, 101–102
online shopping, 118
of relative of addict, 229–230
screen use affecting, 75
stress
always-on readiness and, 166
decreasing, 203
disrupting sleep, 167
ease of access and, 202
hypertension and, 163
information causing, 92
pornography and, 142
as symptom, 165–166
technology and, 87–88
symptoms, 153
video games, 96, 99, 111–112, 149
endless choice, 87–88, 118, 130, 134
entertainment
anticipation and, 63–64
digital devices and, 62–63
infotainment, 64
instant gratification, 63–64
intoxication from, 62
limiting specific, 207–208
synergistic amplification of, 61–64
escapism, 50, 104, 108
e-sports, 109–110
expendable activity, 94
extinction resistance, 77
eye movement desensitization reprocessing (EMDR), 242–243
eye strain, 162

F

family support groups and therapy, 228–231
fear of missing out (FOMO), 201, 257, 294

nodes of Ranvier, 32

NOMOPHOBIA (fear of no mobile phone), 201, 257

Norcross, John C., 236

Norton Family, 221

notifications
  defined, 13–14
  increasing addictive potential, 148
  removing, 210–211, 256–257
  turning off, 299, 312

nucleus accumbens
  defined, 30
  in developing brain, 44–45
  dopamine during sex, 128
  function of, 28
  up-regulation and down-regulation, 34–35

O

obesity, 160

obsessive-compulsive disorders (OCD), 228, 246

old brain, 29

online gambling
  addiction to, 120–122
  as behavioral addiction, 10
  investing as, 124
  maybe factor and, 120
  medication for, 227
  overview, 16
  self-assessment test, 188–189
  study of, 60
  in video games, 110

On-line Gamers Anonymous, 220

online investing, 17, 123–124, 188–189

Online Pornography Test, 190–192

online sex. *See* pornography

online shopping
  addiction to, 117–120
  overview, 17
  self-assessment test, 188–189
  in video games, 110, 118

outpatient and partial hospital programs, 239

outpatient counseling and therapy, 238

overt sexuality, 139–140

P

paraphilia, 138

parenting. *See also* relationships
  blacklists and, 207, 272–273
  Child Technology Test, 194–197
  communication, 268–269, 302
  contributing to addiction, 152
  digital detoxes and tech reboots, 270–271
  establishing low-tech days, 270
  family therapy, 230–231
  guidelines for, 266–269
  limiting specific entertainment, 207
  maintaining consistency, 273–274
  monitoring screen use, 272–274
  overview, 49–51, 265–266
  removing technology, 205
  resources, 219–222
  setting example, 274–277
  setting limits
    communication, 304–305
    as family, 267–268
    overview, 269–272
  Tech Sabbaths, 271–272
  whitelists and, 207, 272–273

partial hospital program (PHP), 239

pay-to-play games, 110

peep shows, 129

Pew Research Center, 52, 82, 202, 221

phantom vibrations, 161–162

phishing, 114–117

physical symptoms, 158–168
  DVTs and blood clots, 160–161
  elevated cortisol levels, 165–166
  erectile dysfunction, 159
  eye strain, 162
  gray matter changing in brain, 162–163
  hypertension, 163
  obesity, 160
  repetitive motion injuries, 161–162
  sleep and circadian rhythm disruption, 166–168

physiological dependence, 39–40

PIMU (Problematic Interactive Media Use), 219

psychotherapies
  CBT, 225–226
  family support groups and therapy, 228–231
  group therapies, 231–232
  MAT, 226–228
  overview, 225

Q

Qustodio app, 206, 220

R

Rahming, Melissa, 220
ransomware, 114–117
RaS (reticular activating system), 149
rationalizing, 151–152, 253, 256
Real-Time 100 Living Plan
  creating list, 314
  future challenges and, 257–258
  overview, 211–214
receptors, 30, 34–35
recovery process, 235
reflected echoes, 48
reflected self-esteem, 77
relapses
  defined, 217
  depression and, 245
  ease of access and, 241
  overview, 38–39
  as stage of readiness, 236
  treatment and, 235
  triggers and, 211
relationships. *See also* communication;
    parenting
  additions impacting, 41
  children, 53–54
  considering level of help, 302–303
  dating sites and, 134–135
  denial and, 303
  Internet use impacting, 50–51, 66–67
  lies and, 306

not giving up on loved one, 307
reducing conflicts in, 277
social, 74–75
social media and, 81–82
technology impacting, 154, 252
teenagers, 53–54
validation and, 65
video games and, 99–100, 104–105
re-regulation, 34–35
research-based data, 223–224
residential treatment centers, 239–240
resources for IAD, 218–222
Restart Recovery, 220
resumption as stage of readiness, 236
reticular activating system (RaS), 149
reuptake, 33
rewards
  deficiency syndrome
    depression and dopamine, 228
    overview, 157–158
    research on, 164–165
    up-regulation and down-regulation, 34–35
    video games and, 99
  disruption of system in brain, 68, 253–254
role playing, 103, 108, 137–138

S

saltatory conduction, 32
SAMHSA National Helpline, 222
scams
  overview, 114–117
  sex and, 132–133
Science Daily, 175
Science Direct, 222
Scientific Reports, 165
screen drunk phenomenon, 254–255
Screen Time website, 222
screen use
  addiction and, 11–12
  affecting emotions, 75, 156

work productivity,
175–176
World Health Organization (WHO)
gaming disorder and, 59–60, 97
manual by, 147
website for, 222
wrist injuries, 161–162

Y

Young, Kimberly, 60, 219

Z

Zeigarnik effect, 89
Zuckerberg Media, 222

About the Author

Dr. David Greenfield is the founder and clinical director of The Center for Internet and Technology Addiction and former Assistant Clinical Professor of Psychiatry at the University of Connecticut, School of Medicine, where he taught courses on Internet Addiction as well as Sexual Medicine and supervised residents in psychiatry. He is also the consulting medical director at the Greenfield Pathway for Video Game and Technology Addiction at Lifeskills South Florida.

Dr. Greenfield is a leading authority on Internet and technology addiction, and the author of numerous articles and book chapters as well as the book *Virtual Addiction,* which in 1999 rang an early warning regarding the world's growing Internet addiction problem. His recent work is focused on the neurobiology and treatment of compulsive Internet, smartphone, and screen use. He is credited with popularizing the variable reinforcement, or *slot machine,* model of behavioral addictions and the dopamine–behavioral addiction connection.

Dr. Greenfield participates in public, psychiatric, and addiction medicine communities by sharing his expertise through lectures, research, and popular presentations. He has advanced training and board certifications in addiction medicine, clinical psychology, and psychopharmacology. He is a past president of the Connecticut Psychological Association; sits on the board of the Connecticut Association for the Treatment of Sexual Offenders (CATSO); and is a member of the Sexual Medicine Society of North America, the American Society of Addiction Medicine, and the American Psychological Association.

Dr. Greenfield holds a PhD in psychology from Texas Tech University, an MA in counseling from New York University, and an MS in clinical psychopharmacology from Fairleigh Dickinson University, and he received his bachelor's degree with honors in psychology from Ramapo College of New Jersey. He completed his internship and residency at McGuire VA Medical Center, Virginia Commonwealth University, and Fairfield Hills Psychiatric Hospital. Dr. Greenfield has his board certification in Clinical Psychology from the American Board of Professional Psychology and is certified as a NAADAC Master Addiction Counselor. For the last 35 years, he has practiced in Connecticut and maintains licensure in New York, Massachusetts, and Florida.

Dr. Greenfield lectures and consults throughout the United States and abroad, and has appeared on CNN, HBO, *Good Morning America, The Today Show,* Fox News, ESPN, NPR, and the *Dr. Oz Show.* He has also been featured in *U.S. News and World Report, Newsweek, People, Time,* the *Washington Post,* the *Wall Street Journal,* and *The Economist.*

Dedication

I dedicate this book to my children, Joshua and Jonathan, and to all our children, who will be charged with stewarding these miraculous technologies to enhance humanity and keeping sacred the most important thing in life — human connection.

And to the loving memory of my mother, Thelma Lowenthal-Greenfield, who taught me to appreciate the pleasure of words — and for her belief in me when it was not so easy to do so. To my father, Stanley Greenfield, who reminds me of the importance of fatherhood, and to my siblings, Alec, Barrie, and Laurie, and their spouses, Ellen, Steven, and Yale, who continue to bless me as my family. And to my friend and the mother of our children, Marci Korwin.

And to my patients at The Healing Center, The Center for Internet and Technology Addiction, and Greenfield Pathway, who teach me about the courage to change, and who give me the gift of helping them heal.

Author's Acknowledgments

This book would not be complete without the acknowledgment of many who have been with me on this journey. Special thanks to Sheila Hageman, MFA, who believed in me and my writing and who would remind me that, yes, I am a writer, and for her technical guidance, expertise, and loving support. Thanks to my office manager Tracey R. Fernandes, who allowed me the pursuit of writing by her taking care of everything else. Deep thanks to Dr. Clifford Sussman for his enthusiastic technical and scientific review of this text. Final thanks for their spot-on editing and editorial assistance to Executive Editor Lindsay Lefevere, Managing Editor Michelle Hacker, Project Manager and Development Editor Georgette Beatty, and Copy Editor Marylouise Wiack.

When I began my career focusing on Internet and technology addiction in the mid-1990s, little had been published on the subject, and we were left to navigate the new waters of this nascent disorder without a map. My 1999 study publication, *Psychological Characteristics of Compulsive Internet Use: A Preliminary Analysis,* represented early research that was attempting to examine the addictive nature of the Internet when we knew very little about it. I thank all the researchers, scientists, and clinicians in the years before, during, and since, who have been an influence on my work, and who have continued to provide inspiration in the specialty of the behavioral addiction to the Internet.

The digital world has changed dramatically since my early research, writing, and clinical work on this emerging issue, but I would be remiss if I did not specifically acknowledge those who were so impactful early in my work. To the memory of my friend and colleague, the late Dr. Alvin Cooper, for his early adoption of the importance of the Internet in impacting human sexuality and sexual behavior, and who sparked my work in publishing on this subject; his *Triple-A Engine* concept was integral in my work. In memory of Dr. Kimberly Young, whose pioneering early pilot study lit the pilot light of my own research and my first book, *Virtual Addiction*. And to the prolific work of my University of Connecticut, School of Medicine colleague, Dr. Nancy Petry, whose rigorous scientific study of video game addiction helped advance this field; her passing was a great loss. Lastly, to the late Dr. Maressa Hecht Orzack; without her, our early publications on Internet sexual behavior, and my work on sex and the Internet, would not have been possible.

I am grateful to the many people whose clinical and scientific work continues to motivate, influence, and inspire me in the now-maturing field of Internet and technology use and addiction.

Publisher's Acknowledgments

Executive Editor: Lindsay Sandman Lefevere

Managing Editor: Michelle Hacker

Project Manager and Development Editor:
Georgette Beatty

Copy Editor: Marylouise Wiack

Technical Editor: Clifford Sussman, MD

Proofreader: Debbye Butler

Production Editor: Mohammed Zafar Ali

Cover Image: © DisobeyArt/Shutterstock

Printed and bound by CPI Group (UK) Ltd, Croydon, CR0 4YY

27/10/2024

14580323-0001